HORTICULTURE

About the Authors

Dr. J. Auxcilia is Ph.D in Horticulture from TNAU, Coimbatore from 1988-1998. She was the recipient of ICAR Junior merit fellowship and awarded with two gold medals and one prize for Ph.D (Hort.). Totally she won 16 awards/prizes as a merit of her research and education activities. Handled more than 35 UG courses, 10 PG and Ph.Dcourses and guided 7 students as Chairperson at PG and Ph.D level. She has published 35 research papers, two books including a catalogue on papaya, two e-books and two e-guides.

She has expertise on Papaya and Banana. Germplasm curator for papaya and banana and involved in resistant breeding programme. Acted as National level lead scientist for papaya projects under ICAR- AICRP (Fruits) programme.

Released varieties TNAU Co. 8 papaya, Co. 2 banana, CoTH.1 tomato, COBH. 1 brinjal, CoBgOH. 1 bitter gourd, as one of the scientists. Developed DUS guidelines for papaya (PPV & FRA). Contributed for technologies *viz.*, bunch cover for banana,stem injector for banana, bud pollination technique in tomato *etc.*

Dr. Veena Amarnath is Ph.D in Horticulture from G.B.Pant University of Agriculture and Technology, Udham Singh Nagar, Uttarakhand. She has a vast experience in teaching, research and extension. She worked in TNAU and was one of the scientists responsible in release of bhendi hybrid COBH.1 and trained and empowered about 200 farm women in hybrid bhendi seed production with the help of DST- GOI funded project.

Dr. Amarnath is author of a book on 'Nursery and Landscaping'. She has written a chapter for a book on 'Vegetable Science'. She has 17 international and national publications to her credit. Presently she is a Horticulture Consultant.

Dr. S. Parthiban pursued his B.Sc (Agri.) in Annamalai University, Chithambaram. M.Sc(Hort.) and Ph.D at HC & RI, TNAU, Coimbatore from 1886 to 1998. He has contributed for release of the varieties *viz.,* Sapota- PKM - 4 (SA), Sapota – PKM – 5 (SA), PKM-1 Manila Tamarind (MT) and PKM - 1 Davana. He developed a Technology for "High Density Planting" in Mango referred as "Double Hedge row planting system" for mechanized cultivation. Handled more than 63 UG and PG courses. Assembled 14 types of avocado, 21 types of jack fruit, 150 varieties of grapes and new-ecotypes of panneer grapes *viz.*, Chinthamani, and introduced new improved and high yielding grape varieties *viz.,* Red Globe, Medika and A18-3.

Dr. V.P. Santhi, pursued her B.Sc. (Hort.), M.Sc. (Hort.) and Ph.D at TNAU from 1987 to 2004. She was the recipient of ICAR Junior merit fellowship for M.Sc. (Hort.) and awardee of CSIR-Senior Research Fellowship for Ph.D Hort. Totally she won 7 awards as a merit of her research and education activities. She was specialized on fruit crops especially on Banana, Guava, Pomegranate and Pear and temperate vegetables on French beans, Garlic and Carrot. Handled more than 53 UG courses.The varieties released are French bean – Ooty -3, Garlic Ooty 2 and Butter Pear Ooty -1. Assembled 72 types of garlic, 36 types of French bean, 32 guava and 18 pomegranate in germplasm for further conservation.

Dr. K. Indhumathi, pursued her B.Sc.(Agri.), M.Sc.(Hort.) and Ph.D at TNAU, Coimbatore from 1996-2009. She was the recipient of TNAU Merit scholarship for B.Sc.(Agri.), KIADEF fellowship for M.Sc. (Hort.) and ICMR fellowship for Ph.D. She has specialized in Floriculture and Medicinal crops. She has published 15 research papers, 14 book chapters and two books. She has handled two projects on Nutrition garden demonstration in tribal areas of Dharmapuri district.

Dr. L. Pugalendhi, is Ph.D in Horticulture from TNAU, Coimbatore, recipient of more than 40 externally funded projects worth of Rs.500 lakhs. Assembled 800 germplasm accessions of vegetable crops. Developed 19 horticulture crop varieties/hybrids in vegetable crops and a few others. To his merit, developed brinjal grafting technique for the first time in India to mitigate soil borne pathogens, besides grafting techniques in bitter gourd and cucumber. Developed technologies for protected cultivation of tomato, capsicum *etc.* Recipient of Lasam Amit Singh Memorial award for outstanding contributions in AICRP Vegetable Crops during 2010. He has handled UG, PG and Ph.D courses for more than 15 years including student's guidance as Chairman.

Dr. M.S. Aneesa Rani is Ph.D in Horticulture, pursued her BSc. (Hort.), M.Sc. (Hort.) and Ph.D from TNAU during 1988 to 2002. She handled 21 UG and 4 PG courses. Recipient of ICAR Junior Merit Fellowship and the prestigious Tamil Nadu State Best Women Scientist Award in 2010, with a total of 8 awards/prizes. A voluminous work on Cashewnut resulted in the development of first F_1 hybrid in Cashew. She has contributed for development of five varieties including horticulture and other crops and developed high density planting system for cashew. Published 20 research papers, 5 books, 19 book chapters. Obtained many projects on cashew and Cocoa. Served to the farming community through KVK activities as Professor and Head of Regional Research Station and Krishi Vigyan Kendra, Vridhachalamat TNAU.

HORTICULTURE
Principles and Practices

Auxilia James
Associate Professor (Hort.)
Department of Fruit Science, Horticultural College and Research Institute
Tamil Nadu Agricultural University, Coimbatore – 641 003

Veena Amarnath
Professor (Hort.)
Department of Fruit Science, Horticultural College and Research Institute
Tamil Nadu Agricultural University, Periyakulam – 625 604

S. Parthiban
Horticulture Consultant, Coimbatore – 641 003, Tamil Nadu

V.P. Santhi
Assistant Professor (Hort.)
Department of Fruit Science, Horticultural College and Research Institute (Women)
Tamil Nadu Agricultural University, Trichy – 620 027

K. Indhumathi
Assistant Professor (Hort.)
Department of Fruit Science, Horticultural College and Research Institute (Women)
Tamil Nadu Agricultural University, Trichy – 620 027

L. Pugalendhi
Dean (Horticulture), Horticultural College and Research Institute
Tamil Nadu Agricultural University, Coimbatore – 641 003

M.S. Aneesa Rani
Professor and Head
Department of Fruit Science Horticultural College and Research Institute
Tamil Nadu Agricultural University, Coimbatore – 641 003

nipa®

NEW INDIA PUBLISHING AGENCY
New Delhi – 110 034

Citation: *Auxcilia, J., Veena Amarnath, Parthiban, S., Santhi, V.P., Indhumathi, K., Pugalendhi, L. and Aneesa Rani, M.S. Horticutlure: Principles and Practices, New India Publishing Agency, New Delhi*

NEW INDIA PUBLISHING AGENCY
101, Vikas Surya Plaza, CU Block, LSC Market
Pitam Pura, New Delhi - 110 034, India
Email: info@nipabooks.com
Web: www.nipabooks.com

For customer assistance, please contact
Phone: + 91-11-27 34 17 17 Fax: + 91-11- 27 34 16 16
E-mail: feedbacks@nipabooks.com

ISBN : 978-93-89571-99-8

Composed and Designed by NIPA.

Preface

Horticulture encompasses science, technology and aesthetics of producing fruits, vegetables, flowers, medicinal and aromatic crops. Horticultural crops require intensive care right from planting through cultural operations, management, harvesting, packing, marketing, storage, value addition and processing. To flourish as a professional horticulturist, one needs to have a comprehensive and proficiency of the fundamentals and technologies in horticulture.

This book is a meticulous presentation of fundamental principles of Horticulture and it gradually transcends and culminates to advanced technologies.It will be a ready reckoner, which will be useful to provide ready tips to any undergraduate faculty. Chapter 1 to chapter 4 give a broad description of fundamentals in horticultural sciences and the institutes involved in furthering the interest of horticulture.Chapters 5 to chapter 14 explain about the tools, inputs and techniques used in crop management. Chapters 16 to chapter 19 deal with use of growth regulators, protected cultivation, good agricultural practices andorganic horticulture. With the increase in pollution, people are keen to produce their own vegetables and chapter 20 focuses on that need of people. Chapter 21 provides an insight to applied biotechnology related to horticulture *viz.,* micropropagation and genetic engineering. The chapters 22 and 23 enlighten about the scope of aesthetic aspects of horticulture. Horticulture therapy is a novel concept to deal with present day health issues and this is explained in chapter 24. Climate change and irresponsible human behavior has caused stress conditions for crop growth and it is adequately dealt in chapter 25. It is estimated that at least 20 – 30% of the horticulture produce is wasted as post-harvest loss and hence chapter 26 and 27 are designed to provide post-harvest handling and value addition.

A complete treatise as the present one cannot be completed alone. The author wish to thank and acknowledge the contribution of all the authors. The entire team is indebted to **Dr. N. Kumar, Vice Chancellor** of Tamil Nadu Agricultural University for his inspiring and motivating support throughout the preparation of this book.The Vice-Chancellor himself has written quality resource material for horticulture fraternity.Our sincere thanks to **Dr. P. Masilamani**, the then **Dean, HC & RI (W), Trichy**, but for his encouragement and guidance, this book would not have manifested into a complete and quality document. We also thank and acknowledge the support of staff members of HC & RI (W), Trichy and HC & RI, Coimbatore.

In recent years, a number of developments have been witnessed in the field of horticulture, their nature and scope need thorough scientific understanding. It goes without saying that, whatever be the professional study, a better understanding of the basics is the first challenge in the growth of a student in to a professional. This book will provide the necessary impetus, knowledge and understanding of horticulture as a whole, including the technologies that help to deal with the problems that exist in the new millennium.

Authors

Contents

1

Genesis and Global Scenario of Horticulture

Horticulture deals with the art, science, business and technology of cultivation and management of horticultural plants. The word Horticulture is derived from two Latin words viz. 'Hortus' (means garden) and 'Culture' (means growing). Horticulture is an aesthetic science that deals with cultivation of fruit, vegetable, spice, plantation, medicinal, aromatic and flower crops. There is a vast therapeutic use of horticulture wherein gardening and aroma therapy are used to deal with new age maladies like depression and various addictions.

In ancient times, huge gardening spaces to grow large number of fruits, vegetables and flowers were available and these areas were demarcated with the help of high walls. This enclosed area was referred to as garden. This led to the term garden culture. The land area is decreasing due to industrialization and population growth, at the same time, there is a drastic increase in demand for horticultural produce. To meet the increased demand for horticultural produce and to cope up with the space constraints, multitier cropping systems, vertical gardening and other intensive cultivation practices are being resorted to. Hence, horticulture may also be described as the "Practice of growing and manipulating plants in a relatively intensive manner".

Presently, fruits, vegetables, flowers and ornamentals are grown not only within the home premises, but also in large quantities on a commercial scale. In other words, many horticultural crops throughout the world are grown primarily as business enterprises. Thus, horticulture has two aspects in general i.e., the amateur and the professional. In addition, with the establishment of horticultural experiment stations as an integral part of colleges and universities, comparatively a large amount of scientific information on the behaviour and cultivation of horticultural plants and products is freely available. This information or literature, constitutes the science of horticulture, and the application of this scientific information constitutes the "Technology of Horticulture." Thus, present day horticulture may be defined as the science and technology of production, processing and merchandising of fruits, vegetables, flowers and ornamentals where each crop has its own science and technology. Consequently, horticulture may appear differently to different people. To a teacher, professor, research scientist, extension official and landscape architect, it is an applied science and a profession, while to a commercial producer, wholesaler, a retailer of horticultural products, or to the manufacturers or seller of horticultural products, it is a business. It is also a fascinating hobby for connoisscur.

Genesis of Horticulture

The word 'Horticulture' was first mentioned by E. Philips in the New World of English Words, London, 1678. Horticulture, though considered to be part of Agriculture, is a prime agricultural practice followed by primitive man. Horticulture is as old as Adam and Eve. Many trees have references in ancient epics and folklores. According to the Bible, the inception of humanity lay in Eve's offer of the Forbidden fruit to Adam in the Garden of Eden. The cultivation of plants started primarily for human consumption. In the days of yore, saints and sages maintained their livelihood by eating fruits alone. Thus, it justifies that Horticulture is an age old prime agricultural practice. In Valmiki's Ramayana, it was mentioned that Rama had spent his first day of vanavasa under the tree Thopasatharu (*Balanites roxburghii*). The Asokavana is described as a garden of asoka trees (*Saraca indica*) at Sri Lanka. In Sabhaparva of the Mahabharata by sage Vyasa, a full description about ornamental and fruit trees viz., asoka *(Saraca indica)* champak (*Michelia champaca*), nag-champa (*Mesua ferrea*), Sal (*Shorea robusta*), mango (*Mangifera indica*) etc. has been given. These plants were planted at Indraprastha (the capital of the Pandavas). Even before the existence of Buddha (567 B.C), the pipal (*Ficus religiosa*) was regarded as a sacred tree. Gautama Buddha, received enlightenment under bodhi tree (*Ficus religiosa*). Kalidasa during 5th century A.D, described the asoka tree in most of his plays, especially mentioning the arjuna tree (*Terminalia arjuna*), vakula (*Mimus opselengi*), parijatham (*Nyctanthes arbortristis*) and kavidhara (*Bauhinia variegata*) among the beautiful flowering plants.

Moghal emperors like Babar, Akbar, Jehangir, Shah Jahan and others made beautiful gardens at Agra, Delhi, the Kashmir Valley, Lahore, Pinjore and Punjab during their reign in India. They grew a variety of mangoes in their gardens and many of the mango cultivars are still under cultivation like himayudhin, khader and jehangir, to name a few. Mango, the king of fruits was previously called as Amra-Phal. It is also mentioned in early Vedic literature as rasala and sahakara. Mango finds reference in Brihadaranyaka Upanishad and the Puranas.It has religious importance in Jainism because the Jain goddess Ambika is traditionally represented as sitting under a mango tree. Many flowers find reference in

Urdu Poetry. Tulips and narcissus are found carved in marble on the walls, floors and ceilings in Red Fort at Delhi, and the Taj Mahal at Agra and many more architectures. Fruits like fig, pomegranate, grapes, olives are mentioned in the Bible also. Fruits, vegetables and medicinal plants form an integral part of the ancient ayurvedic system of medicine and chinese herbal medicine system. One of the Seven Wonders of the World, the hanging gardens of Babylon (660 B.C), is a good example of multi-tiered garden containing a wide variety of gorgeous flowering trees, shrubs and vines. It is called hanging garden because the trees are planted on raised terraces. Planting of flowers and trees has been followed by people in China and Japan. In Japan, the "bonsai" is given as ancestral property from one generation to the other. Thus, the science of horticulture prevailed and developed since ancient times.

Nature of horticultural crops

Garden crops traditionally include fruits, vegetables, ornamental plants, spices, plantation crops, medicinal and aromatic plants etc. Horticulture differs from Agriculture. Horticulture crops require intensive care in planting, carrying out cultural operations, manipulating growth, harvesting, packing, marketing, storage and processing. Many horticultural crops especially fruits and vegetables are consumed in raw state, and are highly perishable due to high moisture content. On the contrary, agricultural crops like rice, wheat etc., can be stored for a long time as the economic part is mostly the endosperm.

Before "Horticulture" was given the status of a full-fledged subject, it was taught by botanist and this aspect of botany was called as "Applied Botany". Horticulture is a vast subject comprising of the knowledge on taxonomy, anatomy, pathology, physiology, entomology, genetics and biotechnology as well as postharvest and processing of the horticultural crops. Hence, now, it may be termed as "Horticultural Science". In present days, botany deals with all crops, while horticulture deals with fruits, vegetables, spices and plantation crops, medicinal, aromatic and ornamental crops. Hence, a Horticulture scientist works in coordination with all other specialists like plant physiologist, pathologist, bio-chemist etc.

Importance of Horticulture

- Horticultural crops are important to all of us as a source of food. Around 40 per cent of the food we consume daily consists of horticultural crops. The fruits with which we start our meal, the salads viz., onion, tomato, carrot, radish and processed products like jelly or jam and pickles, all provide calories to the diet, besides vitamins and minerals. We also depend largely on vegetable proteins found in peas and beans. The aesthetic value of flowers in home gardens, shade trees along streets, parks and pleasing landscape designs are the real value of horticulture in our day to day life. Thus, horticultural crops are an integral part of our lives as a **source of food and well-being.**

- Horticultural crops provide complete nutrients, vitamins, minerals and fibres that are required for a healthy body and very aptly, fruits and vegetables are regarded as "protective foods". Horticultural crops offer food security as they form a balanced diet. Unlike rice and cereals which majorly contain carbohydrates and proteins, horticultural crops have all the essential nutrients required for our body like carbohydrates, proteins, fats, fibre, vitamins and minerals. Banana, sweet potato and cassava serve as the staple food in many African countries. Fruits such as banana, avocado, olive etc. are also good sources of energy. Fruits yield more calorific value. An adult normally requires 11,00,000 calories per year. This can be obtained from 0.44 ha of wheat. But the same quantity can be obtained from 0.03 ha of banana or 0.06 ha of mango. Hence, horticultural crops provide **food security**.
- Fruits and vegetables are rich in anti-oxidants, especially vitamin C and carotenoid pigments like beta carotene, lycopene, anthocyanin etc. Anti-oxidants play a major role in free radical scavenging and prevent damage to tissues and safe guard the body against toxins. Anti-oxidants are considered as anti-cancerous. Green tea is considered a good source of antioxidants. Vitamin C is present in large quantities in west indian cherry, amla, guava, papaya, mango, citrus fruits and green leafy vegetables. Several fruits and vegetables such as mango, papaya, pumpkin and greens contain beta carotene, which is a precursor for Vitamin A. Almond, walnut and cashewnut are rich in fat and protein and have high calorific value. Avocado has 24% fat, which is cholesterol free and good for health. Several fruits are rich in minerals. Banana is a good source of calcium and potassium. Wood apple is a good source of Iron and calcium. Amla, guava and papaya are rich in pectin. Hence, horticultural crops provide **nutritional security.**
- Horticultural crops are high yielding as compared to agricultural crops and hence highly remunerative. Spices, medicinal, aromatic and flower crops are export oriented and high value crops with intensive cultivation, horticultural crops are capable of giving more output per unit area. They give higher returns per unit area in terms of energy, money, job, etc. Banana gives an average yield of 40-60 t/ha, papaya 150 to 200 t/ha and potato 20-25 t/ha whereas paddy gives 5-6 t/ha only. Cultivation of horticultural crops can be taken up as a hobby via kitchen and rooftop garden to supplement family income. Thus horticultural crops are important for **income generation.**
- Horticultural plants and their products, directly or indirectly constitute the raw material for many industries like plantation crops, fruit crops, medicinal

plants, aromatic plants, spices and ornamental crops. Horticultural crops are labour intensive. These crops require special cultural operations such as training, pruning, thinning, harvesting, grading etc., at various stages of growth and development and hence the man-days requirement is also high. Fruit crops require 860 man days/annum ha as against 143 man days/annum ha for cereal crops whereas the crops like grapes, banana and pineapple need 1000-2500 man days/annum ha. Processing industries and value added products are the major avenues to provide employment to the rural and urban sectors in India. Cashew processing industry alone provides employment to over 5.5 lakh people annually. Tea and coffee processing industries provide employment opportunity to many more. Horticultural crops provide material for processing and preservation industry. Jam, jelly, marmalade, juice, squash, crush, wine, pickles, canned vegetables, dried vegetables, dried flowers, dried herbs, chips and wafers are popular. Thus, horticultural crops help in **employment generation.**

- Horticulture crops generate substantial national revenue. As a foreign exchange earner, the contribution of horticultural crops is more than any other agriculture crop. Mango, banana, grapes, papaya, onion, potato, roses, jasmine, cashewnut, tea, coffee, coconut, arecanut, black pepper, cardamom, ginger, turmeric, chillies etc. have export value. European and gulf countries are the major importers of horticultural produces. Thus, horticultural crops help in earning **foreign exchange.**
- Horticultural crops are very large in number; hence choices are available to suit any kind of climatic conditions and soil quality. It can be grown in almost any kind of soil. Perennial horticultural crops along with forestry crops can protect the environment by providing vegetative cover on soil, thus preventing the loss of soil due to erosion by wind and water. Trees like *Polyalthia, Thespesia populina and Nerium* are efficient in pollution control by collecting dust. *Solanum melongena* and *Cyamopsis tetragonaloba* effectively remove sulphur dioxide from the atmosphere. Shrubs and hedges check noise pollution. Growing crops increases aesthetic value and is good for the environment too, since growing plants reduces global warming, regulates rains and minimizes soil erosion. Trees are known to help collect rainwater and transfer it to the underground thus enhancing the groundwater table too. They also provide habitat to conserve biodiversity. Hence, horticultural crops ensure **environment safety.**
- Horticulture crops are important also as a source of medicines, flavour, aroma, alkaloids, oleoresins, nuts and some oils. For example, bitter gourd juice, jamun seeds, flowers of *Senna auriculata* and fenugreek are recommended for management of diabetes. Medicinal plants like senna is a great laxative Keelanchi *(Phychanthus niruri)* for jaundice, *Gloriosa*

superba for arthritis, periwinkle *(Cartharanthus roseus)* to fight cancer, aswagantha *(Withania somnifera)* for general vitality etc. have been in use from the Vedic periods. The extracts are used in all forms of medicine, including allopathy. Hence, horticultural crops act as health and wellness promoters.

- Horticultural crops are useful for various mind related disorders. Horticultural therapy is the practice of engaging people in plant or gardening activities to improve their bodies, minds and spirits. Research confirms that health benefits accumulate when people connect with nature and plants by viewing and/or interacting with them. Horticulture therapy is a growing profession in rehabilitation and addiction centres, vocational centres, youth mental health programs, hospitals and nursing homes, old age homes or senior citizen centres, mental health centres, schools and colleges. Aroma therapy is another scope for horticultural plants. Aroma from lavender calms the mind. Studies have shown that the gaseous aromatic hydrocarbons released by plants have a strong capacity to disinfect even by infiltration through human skin. Thus when it is absorbed by human skin or mucosa, it can stimulate the production of immune protein and relax the nervous system, finally enhancing the body's resistance. Hence, horticultural crops are useful for **therapeutic purposes.**

Scope of Horticulture in India

1. Incentive for the farmer

Horticultural crops provide more returns in terms of per unit area of production, export value and value addition compared to agricultural crops.

2. Adaptability

India is bestowed with a great variety of climatic and edaphic conditions. Likewise, we have a multitude of soils ranging from loam, alluvial, laterite, medium black, rocky shallow, heavy black, sandy etc. and thus a large number of crops can be accommodated with very high level of adaptability. Wastelands, including coastal lands can be best utilized to cultivate hardy horticultural crops like fruits and medicinal plants. Thus, there is lot of scope for horticultural crops.

3. Necessity

Nutritional security for the people of the country has become a priority. To meet the nutritional requirements in terms of vitamins and minerals,cultivation of horticulture crops should be done in sufficient quantities to provide a bare minimum of 200 g of fruits and 200-250 g of vegetables per head per day, as per World Health Organization (WHO) report.

4. Export value

Fresh fruits like mangoes and grapes, vegetables such as onions and potatoes, flowers like roses, plantation crops cashewnut, tea, coffee, coconut, arecanut, and spice crops like black pepper, cardamom, ginger, turmeric, chillies, etc., constitute the bulk of the export basket. European and *Gulf* countries are major importers of horticultural produce. At present our share in international trade of horticultural commodities is less than one per cent of total trade. Moreover, these commodities (spices, coffee, tea etc.), fetch 10-20 times more foreign exchange per unit weight than cereals. Therefore, taking advantage of globalization of trade, proximity big markets and the volume of production, our country should greatly engage in international trade which would provide scope for growth in this sector.

5. Physical and Mental wellbeing

Horticulture therapy and aroma therapy provide physical and mental wellbeing to the person practicing it, hence horticulture is important for a mentally and physically fit and healthy society.

Global and National Scenario of Horticulture

Horticulture sector comprises of a wide variety of crops. Horticultural crops include fruits, vegetables, spices, plantation, flowers, medicinal and aromatic crops. Horticulture sectors is one of the major drivers in agriculture sector growth in India.

The need for diversification in horticulture sector was acknowledged by the Government of India in mid-eighties by focusing its attention on investment in this sector. Presently horticulture has established its credibility in improving income through increased productivity, generating employment and enhancing global presence.

Recently, Horticulture has taken great strides from a rural to commercial venture. The Department of Agriculture and Co-operation (DAC) of the Ministry of Agriculture is the nodal department for overviewing horticulture development in the country. It implements diverse programmes through Department of Horticulture in all the States and provides the leadership to coordinate activities for the promotion of horticulture. However, efficient implementation of programme depends upon timely availability of statistics. The exponential growth of horticulture sector in India is very encouraging. Out of the total agricultural production in India, horticulture out put is around 33% which is a large share.

Global Scenario

India is the second largest producer of vegetables and fruits and our presence in global market is prominent. A variety of horticultural commodities are exported from India, among which grapes occupies the premier position in exports with 188.2 thousand tonnes valued at Rs.1,89,994 lakhs followed by banana and mango. Export of fresh vegetables like peas and potatoes also witnessed growth. Market fluctuation is found to be maximum for the vegetable crops like onion, potato and tomato.

National Scenario

India has perceived voluminous increase in horticulture production over the last few years. Signicant progress has been made in area expansion resulting in higher production. Over the last decade, the area under horticulture grew by about 3% per annum and annual production increased by 5.4%.

Area : 25.43 million hectares

Production : 311.71 million tonnes

Productivity : 12.25 tonnes/ha

Among horticultural crops, vegetables have shown a tremendous increase in area and production. The production of vegetables has increased from 58.5 million tonnes in 1991-92 to 184.39 million tonnes in 2016-17.

In horticultural production, share of fruits is 31.2%, vegetables 59.2%, flower and aromatic crops 1.2%, plantation 5.8% and spices 2.6%. In the total horticulture production, 90% comprises of fruits and vegetables. From the total export of floriculture products from India, the major share (20%) goes to United States. The exported spices and spice products are chilli, mint products, spice oil, oleroresins, cumin, turmeric, pepper, curry powders, cardamom seeds and other spices like tamarind and asafoetida. India ranks 3rd in volume and 2nd in value of the world trade in aromatic oils.

Major distribution of crops in the country

Nature of crop	Crop	State
Fruit crops	Banana	Tamil Nadu, Andhra Pradesh, Maharashtra, Gujarat, Karnataka
	Mango	UP, Gujarat, Karnataka, Andhra, Bihar
	Citrus	Andhra Pradesh, Maharashtra, Punjab, MP, Gujarat
Vegetables	Potato	UP, Bihar, WB, Gujarat, Punjab
	Onion	Maharashtra, Gujarat, MP, Karnataka, Bihar
	Tomato	Andhra Pradesh, Odisha, Karnataka, MP, WB, Bihar
Spice	Chilli	Andhra Pradesh, Karnataka, WB, MP, Odisha

Area, Production and Productivity of Horticultural Crops in India (2016-17)

S.No.	Crops	Area (M. Ha)	Production (M. MT)	Productivity (MT/Ha)
1.	Fruits	6.50	97.35	14.96
2.	Vegetables	10.26	184.39	17.97
3.	Spices & Condiments	3.87	8.10	2.09
4.	Plantation crops	3.74	18.08	4.83
5.	Flowers, medicinal and aromatic plants	1.40	3.65	3.49

Fruits

- India ranks second in fruit production next to china in both area and production
- It accounts for about 10.50 % of the world's fruit production.
- Per capita availability of fruits is 207.7 g /person/day.
- India ranks first in production of papaya, mango, mangosteen, guava, banana, lemons and lime.
- Major fruit producing states are Andhra Pradesh (15.6%) followed by Maharashtra (12%), Uttar Pradesh (10.8%) and Gujarat (9.2%).
- The total fruit production was the highest in Andhra Pradesh (120.98 lakh tonnes) followed by Maharashtra (103.78 lakh tonnes).

Vegetables

- India is the second largest producer of vegetables next to China in both area and production.
- It accounts for about 11% of the world's production of vegetables.
- Per capita availability of vegetables is 393.8 g /person/day.
- It ranks first in okra production and ranks second in production of tomato, potato, onion, brinjal, cabbage, cauliflower and broccoli.
- The total vegetable production was the highest in Uttar Pradesh (15.4 million tonnes) followed by West Bengal (15.03 million tonnes) and Madhya Pradesh (9.52 million tonnes).

Spices

- India is known as 'Home of Spice' and Kerala is known as 'land of spice'.
- India is the largest producer, consumer and exporter of spices in the world. There are 75 spices grown in the country out of the listed 109.
- The major spices and condiments producing states in terms of revenue are Gujarat followed by Andhra Pradesh and Maharashtra.

Plantation Crops

- Coconut, Coffee, Tea, Cocoa, Rubber, Cashew and Arecanut are the major plantation crops grown in India.
- India is the second largest producer of tea next to china.
- India ranks first in coconut production followed by Indonesia and Philippines.
- In India, the largest area under plantation crops is shared by Kerala (33.48%), Karnataka (27.3%) and Tamil Nadu (23.42%).

Flowers

- India's share in global floriculture trade is 0.6%.
- India is the largest producer of loose flowers in the world.
- Largely grown flower crop in India is Jasmine.
- India is the largest producer of Jasmine concrete and tuberose concrete.
- The highest production of flowers was recorded in Tamil Nadu (17.3%), Andhra Pradesh (15.4%) and Karnataka (11.3%).

Medicinal and Aromatic Plants

- The global demand for medicinal and aromatic plants is growing @ 15% per annum.
- About 627 plant drugs are being traded in India.
- About 90% of the medicinal plants are collected from wild source.
- India is the largest producer of Senna, Davana oil and Isabgol and third largest producer of Periwinkle.
- India is the largest producer of Opium latex.
- India is the largest exporter of *Psyllium* and Senna leaves.
- World production of essential oil is dominated by Brazil (40%) followed by USA (20%) and India (15%).

2

Division and Classification of Horticultural Crops

Horticulture is a wide field and includes a great variety and diversity of crops. In general, these crops are classified as follows:

1. Crops grown for the food and/or vitamin and mineral content, for e.g., apple and sweet potato.
2. Crops grown for their beverage properties, for e.g., tea and coffee.
3. Crops grown to beautify the environment, for e.g., the grasses of the well planned lawn.
4. Crops grown to express sentiments, for e.g., cut roses, jasmine.
5. Crops grown for perfume for e.g., rose, jasmine.
6. Crops grown for spices for e.g., pepper, nutmeg.

Divisions of Horticulture

- **Pomology** (Fruit Culture): This division of Horticulture deals with the study of different fruit crops. Cultivation, management and other aspects of fruit crops are covered under this branch e.g. mango, banana, grapes.
- **Olericulture** (Vegetable culture): This division deals with the study of vegetable crops. Vegetables are nutritive food of plant origin which are normally cooked before consumption or eaten raw as salad. e.g. cabbage, tomato, fenugreek.
- **Floriculture and Landscape architecture**: This division of Horticulture covers flower crops and ornamental plants. E.g. roses, asters, lily, cactus, ferns, etc. Sub divisions of floriculture: i) Commercial floriculture (cultivation of economic flowers like rose, jasmine etc), ii) Arboriculture: Growing of trees, iii) Landscape architecture (planning and beautification of home gardens, public area, industries and business establishments) iv) Ornamental gardening (study of various ornamental plants used to decorate indoor and outdoor gardens).

- **Spices and Plantation crops**: Spices and condiments are used as food adjuncts to add aroma in culinary. Cardamom, pepper, clove provide flavour. Condiments are used as food adjuncts to add taste only. e.g. coriander, cumin. The crops like arecanut, coconut, tea, coffee and rubber are known as plantation crops which are grown in extensive scale.
- **Medicinal and Aromatic plants**: The plants yielding alkaloid and steroid principles which have got preventive and curative properties are known as medicinal plants e.g., periwinkle senna. The parts of aromatic plants yield essential oils. Eg. geranium, lemon grass, mint etc.

Classification of Horticultural crops

Classification is done for ease of understanding the crops, so that the cultivation is proper.

I. Classification of fruit crops

A. Classification based on climate adaptability

In this classification, the fruits trees are categorized into three recognized groups.

i) Temperate fruits

Temperate fruit plants are grown only in places where winter is distinctly cold. These crops require exposure to specific chilling temperature for certain period without which they do not produce flowers. These fruit plants are generally deciduous in nature and tolerate frost. Examples are apple, almond, peach, etc.

ii) Tropical fruits

Tropical fruit plants are generally evergreen and are extremely sensitive to cold.They do well under lesser fluctuations of diurnal temperature, light and dark periods. They require a moist warm climate but are capable of withstanding dry weather in some cases. Examples are mango, banana, papaya, sapota, etc.

iii) Sub-tropical fruits

The fruit crops that grow well under climatic conditions which are intermediate between temperate and the tropical climates are known as subtropical fruit crops. They may be either deciduous or evergreen and usually withstand a low temperature but not the frost. Examples are grape, citrus, durian, jackfruit, etc.

B. Classification based on fruit morphology

i) **Simple fruit - Berry:** banana, papaya, grape, sapota and avocado.

ii) **Modified berry-**

- Pepo - water melon.
- Pome - apple, pear, loquat.
- Drupe (Stone) - mango, pear, plum.
- Hesperidium - citrus.
- Nut - cashew, litchi, walnut, rambutan.
- Capsule - aonla, carambola.

iii) **Aggregate fruits-** Custard apple, raspberry, straw berry, black berry

iv) **Multiple fruit-**

- **Syconus** - fig.
- **Sorosis** - jackfruit, pineapple.

C. Based on rate of respiration

i) Climacteric fruits

Fruits which can ripen after being harvested at maturity. They attain a respiration peak even after harvest. They produce a large amount of ethylene. Hence, they can be harvested at full maturity, but even before ripening. Examples are mango, banana, sapota, guava, papaya, apple, fig, peach, pear, plum, annona, tomato.

ii) Non-climacteric fruits

Fruits which do not ripen even after being harvested at maturity. Hence, these fruits should be allowed to ripen on the tree itself. Examples are citrus, grape, pomegranate, pineapple, litchi, ber, jamun, cashew, cucumber, cherry, strawberry.

D. Based on photoperiodic responses

i) **Long day** – Crops require long light period and short nights for flowering are referred as long day plants. These require light for a period of 12 hours or more. Examples are passionfruit, banana, apple.

ii) **Short day** – Crops require short light period and long dark period for flowering are referred as short day plants. These require light for a period of 8-12 hours for flowering. Examples are strawberry, pineapple, coffee.

iii) **Day neutral plant** – Crops in which the flowering is not influenced by day length and hence referred as day neutral plants. Examples are papaya, guava.

E. Based on salt tolerance

i) **Highly tolerant–** Crops which can tolerate salinity level at 8mmhos are classified under this group. E.g., datepalm, ber, amla, guava, coconut.

ii) **Medium tolerant–** Crops which can tolerate salinity level from 3-6 mmhos are classified under this group. E.g., pomegranate, cashew, fig, jamun, phalsa.

ii) **Highly sensitive–** Crops which can tolerate salinity only from 1.5 mmhos to 3 mmhos are classified under this group. E.g., mango, apple, citrus, pear, straw berry.

II. Classification of vegetable crops

A. Classification based on edible part

i) Root vegetables: Examples are carrots, beetroot, radish.

ii) Tuber vegetables: Examples are yam, potato.

iii) Corm: Example is amorphophalus.

iv) Stem vegetable: Example is asparagus.

v) Leafy vegetables:Examples are lettuce, amaranth.

vi) Bulb vegetables: Examples are onion, garlic.

vii) Fruit vegetables: Examples are tomato, okra.

viii) Flower vegetables: Examples are broccoli, cauliflower.

ix) Seed vegetable: Examples are beans, peas.

B. Classification based on botanical families

i) Solanaceae: Examples are tomato, potato, brinjal, chilli.

ii) Cucurbitaceae: Examples are Gourds like pumpkin, ash gourd, bitter gourd, bottle gourd, cucumber, melons.

iii) Alliaceae: Examples are onion, garlic and leek.

C. Based on life cycle

i) Annuals: These plants complete their life cycle, both vegetative and reproductive phases in one season. They have seed to seed life cycle in one season, after which they die. Examples are amaranthus, pumpkin.

ii) Biennials: These plants complete vegetative phase in one season and reproductive phase in another season. Examples are cabbage, cauliflower, beet, radish, carrot.

iii) Perennial: These plants do not die after producing fruits. They remain alive for many seasons, producing fruits every year. Examples are chow chow, asparagus.

D. Based on use

i) Greens–The green leafy part is consumed. Examples are amaranthus.

ii) Salads–These are eaten raw or semi cooked. Examples are carrot, cucumber.

E. Based on climatic adaptation

i) Warm season crops–These crops grow in summer season. Examples are okra, chilli, cucurbits.

ii) Cool season crops–These crops grow in winter season. Examples are cauliflower, broccoli, capsicum.

F. Based on origin

i) Exotic vegetables–These are not native to India. These are introduced from other countries. Examples are cabbage, tomato.

ii) Indigenous vegetables–These are the native to India. Examples are brinjal and amaranthus.

G. Based on growing season

i) Rainy or Kharif season (June-September)

Exampes are brinjal, chili, okra, tomato, cowpea, cluster bean, cucumber, bottle gourd, ribbed gourd, snake gourd, sponge gourd, ivy gourd, ash gourd, sweet potato etc.

ii) Winter or Rabi season (October-January)

Cole crops like radish, carrot, turnip, beet root, garlic, onion, spinach, palak, peas, potato etc.

iii) Summer season (February-June)

Brinjal, tomato, chilli, okra, cowpea, cluster bean, amaranthus, cucurbits etc.

H. Based on salt tolerance

i) **Less tolerant-** peas, beans, radish, potato, brinjal, sweet potato, celery, snake gourd, globe artichoke.

ii) **Moderately tolerant**- onion, carrot, cabbage, cauliflower, broccoli, tomato, musk melon, water melon, bottle gourd, pumpkin, summer squash, amaranthus, chilli, spinach etc.

iii) **Tolerant**- asparagus, palak, turnip, beet root, lettuce, bitter gourd, ash gourd etc.

I. Based on method of culture

Vegetables which require similar method of cultivation practices are grouped here, though they belong to different botanical families. This classification is based on the farmers practice.

Group 1: potato

Group 2: tomato, brinjal, chilli, capsicum (Solanaceous crops)

Group 3: cabbage, cauliflower, knolkhol, sprouting broccoli, Brussels sprout (Cole crops)

Group 4: pumpkin, melon, gourds and squashes (Cucurbits)

Group 5: radish, carrot, turnip, beet root (Root crops)

Group 6: onion, garlic, leek, shallot (Bulb crops)

Group 7: sweet potato

Group 8: okra

Group 9: peas and beans

Group 10: lettuce, celery, parsley (Salad crops)

Group 11: spinach, palak, amaranthus, coriander, fenugreek (Greens)

Group 12: cassava, yam, elephant foot yam, colocasia (Tuber crops)

Group 13: asparagus, globe artichoke, rhubarb (Temperate perennials)

Group 14: drumstick, curry leaf, agathi (Tropical perennials).

III. Classification of ornamental crops

A. Based on growth cycle

i) Annuals–These plants complete their life cycle, both vegetative and reproductive phases in one season. They have seed to seed life cycle in one season, after which they die. Examples are Petunia, zinnia, marigold.

ii) Biennials–These plants complete vegetative phase in one season and reproductive phase in another season. Example is Foxglove.

iii) Perennials–These plants do not die after producing fruits. They remain alive for many seasons, producing fruits every year. Examples are Geranium, coleus, sanseveria.

B. Based on use:

The herbaceous plants are classified as

i) Bedding plants–These are planted in beds. Examples are petunia, zinnia, pansy.

ii) Hanging plants–These plants are used in hanging baskets and rockery. Example is chlorophytum.

iii) House plants–These plants are used for indoor gardening. They require filtered light for best performance. Examples are sanseveria, ficus.

Woody ornamentals plants are classified as

i) Shrubs–Used for shrubbery border, concealing wall or as architectural plant. May be used for topiary also. Examples are hibiscus, pine.

ii) Trees–Trees with bright colour flowers and beautiful foliage are used. The examples are gulmohar, cassia.

iii) Vines–Vines are the creeping plants. Examples are allamanda, jasmine, clematis, vernonia.

IV. Classification of spices

A. Based on life cycle

i) Annual–These plants complete their life cycle, both vegetative and reproductive phases in one season. They have seed to seed life cycle in one season, after which they die. Examples are coriander, fenugreek.

ii) Biennial–These plants complete vegetative phase in one season and reproductive phase in another season. Example is garlic.

iii) Perennial–These plants do not die after producing fruits. They remain alive for many seasons, producing fruits every year. Examples are cardamom, pepper, nutmeg.

B. Based on parts used

i) Seed spices– coriander, fennel and fenugreek.

ii) Flower spices– saffron, clove.

iii) Rhizome– turmeric, ginger.

V. Classification of plantation crops

A. Based on botanical relationship

i) Monocot– arecanut, coconut.

ii) Dicot– tea, coffee, rubber.

B. Based on growth habit

i) Vine– vanilla, pepper.

ii) Shrub– tea, coffee.

iii) Tree– cashewnut.

Nutritive Importance of Horticultural Crops

- Fruits and vegetables play an important role in maintaining a balanced diet.
- They provide energy rich foods, vital protective nutrients/elements and vitamins.
- Comparatively fruits and vegetables are the cheapest source of natural nutritive foods.
- Since most Indians are vegetarians, the incorporation of horticulture produce in daily diet is essential for good health.
- Realizing the worth of fruits and vegetables in human health, Indian Council of Medical Research (ICMR) has recommended consumption of 120g fruits and 280 g vegetables per capita per day. As per World Health Organisation (WHO), 200g of fruits and 200-250g of vegetables is reommended as per capita consumption.
- With the growing awareness and inclination towards vegetarianism worldwide, the horticulture crops are gaining tremendous importance.
- Fruits are a good source of energy. Eg. Avocado, olive etc.
- Fruits like jamun and papaya have certain enzymes which are helpful in metabolic activities leading to proper digestion of food.
- All fruits have one or the other medicinal value.
- They should be eaten in adequate quantity.
- Regular consumption of fruits reduces obesity, maintains health and increases longevity.
- Fruits are attractive in appearance, delicious in taste and easily digestible. Therefore, they are liked by young and old alike.

Functions of fruits and vegetables in human body

- Fruits and vegetables provide palatability and taste.

- They improve appetite and provide fibre for better bowl movement thus helping in keeping good health.
- They neutralize the acids produced during digestion of proteins and fatty acids.
- They improve the general immunity of human body and protect against diseases and deficiencies.
- They are important sources of vitamins and minerals.
- Fruits and vegetables provide higher energy value per unit area compared to cereals.

Nutrients rich in different Horticultural Crops

S.No.	Nutrients, vitamins & minerals	Fruits	Vegetables
1.	Carbohydrate	Mango, Banana, Sapota, Custard apple	Potato, Tapioca, Sweet potato
2.	Protein	Jack fruit, Pomegrante, Custard apple	Peas and Beans
3.	Lipids	Avocado	-
4.	Vitamin A	Mango, Papaya	Carrot
5.	Vitamin - B	Mango, Banana, Papaya, Pineapple	Leafy vegetables
6.	Vitamin - C	Aonla, West Indian cherry, Guava	Cauliflower, Cabbage
7.	Vitamin - D&E	-	Green leafy vegetables
8.	Calcium	Aonla, Custard apple	Onion, peas, Tomato
9.	Phosphorus	Pomegranate, Wood apple	Cucumber, Cauliflower
10.	Potassium	Banana, Apple	Bitter gourd
11.	Iron	Date fruit, Pineapple	Carrot, Bitter gourd, Onion

Role of Horticultural Crops in Human Nutrition

From human nutrition point of view, horticultural crops are the most important to our daily living. Many of the horticulture crops and their products find place in our meals and diet. Human body requires vitamins, minerals, proteins, energy etc. All these are supplied by horticultural crops. Fruits and vegetables are the chief sources of vitamins, minerals, carbohydrates, fats, proteins etc.

Fruits and vegetables are recognized as **protective foods** as they are necessary for the maintenance of human health. Balanced diet relies only on horticultural crops especially fruits and vegetables as it contains all the nutrients requied for human health such as carbohydrates, protein, fat, vitamins and minerals. The bio availability of vitamins and minerals are more from fruits as compared to vegetables.

Vitamins

These are the important constituents of fruits and vegetables and are indispensable part of human diet. Although required in very minute quantities, they are absolutely essential for the maintenance of health. The deficiency of any vitamin from the diet for considerable period may lead to disorder conditions or make the body prone to acquire diseases. Fruits and vegetables supply several vitamins.

Vitamin-A

It is essential for normal growth, reproduction and maintenance of health and vigour. It affords protection against cold and influenza and prevents night blindness. The deficiency of this vitamin results in cessation of growth in young children, night blindness, drying up of tear glands in the eyes, eruption of skin (rashes on the skin) and brittleness of the teeth.

Sources

Fruits-mango, papaya, dates, jackfruit, walnut etc.

Vegetables-Greens like palak, spinach amaranthus, fenugreek, carrot, cabbage, lettuce, peas, tomato etc.

Vitamin B1 (Thiamine)

Tones the nervous system and helps in proper functioning of the digestive tract. Its deficiency in human diet results in beri-beri, paralysis, loss of sensitivity of skin, enlargement of heart, loss of appetite, loss of weight and fall in body temperature.

Sources

Fruits-orange, pineapple, jack fruit, cashewnut, walnut, dry apricot, almond, banana etc.

Vegetables-green chilli, beans, onion, sweet potato, tomato, leaves of colocasia

Vitamin B2 (Riboflavin)

This vitamin is required for body growth and health of the skin.The deficiency of this vitamin causes sore throat, anorexia, cataract, and loss of appetite and body weight and also development of swollen nose.

Sources

Fruits- bael, papaya, litchi, banana, apricot, pomegranate, pear etc.

Vegetables- cabbage, cauliflower, potato, peas and beans, methi, lettuce, asparagus, green chillies, leafy vegetables etc.

Vitamin -C (Ascorbic acid)

This vitamin promotes general health and healthy gums, prevents scurvy disease which is characterized by pain in the joints and swelling of limbs (rheumatism), bleeding of gums, tooth decay and keeps the blood vessels in good condition.

Sources

Fruits: amla, guava, ber, citrus, strawberry, pineapple etc.

Vegetables: tomato, palak, menthi, cabbage, green chillies, spinach, potatoes, peas and beans and carrot etc.

Vitamin-D

This vitamin is necessary for building up of bones, preventing rickets and diseases of teeth.

Sources

All green leafy vegetables are rich in this vitamin.

Vitamin-E

It has an important effect on the generative functions and promotes fertility.

Sources

Green lettuce and other green vegetables.

Vitamin-K

This vitamin prevents blood clotting.

Sources

All green leafy vegetables are rich in vitamin K.

Minerals

Human body requires minerals like phosphorus, calcium, Iron, and Iodine etc. for maintaining good health.

Calcium

It is essential for development of bones regulation of heartbeat, controlling blood clots

Sources

Fruits- acid lime, orange, fig, dried apricots, wood apple etc.

Vegetables- cabbage, greens, beans, carrot, onions, peas, tomatoes, agati, spinach, drumstick leaves etc.

Iron

It is required for production of haemoglobin and it is constituent of red blood corpuscles. Its deficiency causes anaemia, smooth tongue, pale lips, eyes and skin and frequent exhaustion.

Sources

Fruits-custard apple, guava, pineapple, straw berry, grape, black currants, dried dates etc.

Vegetables: carrot, drumstick leaves, beans and agati etc.

Phosphorous

It is essential for maintaining the moisture content of tissues and for development of bones.

Sources

Fruits-guava, grape, jackfruit, passion fruit, orange

Vegetables-carrot, chilli, drumstick leaves, beans, cucumber and onion.

Proteins

These are body building foods. These are essential for growth of the body. The deficiency of proteins in the body causes retarded growth and increases susceptibility to diseases and causes lethargy.

Sources

Fruits- Most of the fruits are low in proteins except guava and banana.

Vegetables-peas and beans are rich in proteins.

Enzymes

These are required for controlling several metabolic activities in the body.

Sources

Papaya-Papain from papaya and bromelin from pineapple.

Fibre and roughages (cellulose and pectin)

Fruits and vegetables supply roughages. These are required for digestion and prevention of constipation.

Sources

Fruits contain low content of fibre. Guava and aonla are better sources compared to other fruits.

Leafy vegetables are rich in fibre content.

Energy foods

Fruits and vegetables contain carbohydrates and fats there by supply energy to human body. Those fruits are banana, dates, apple etc. Nut fruits like walnut, cashew nut and almond etc. supply proteins, besides energy.

Vegetables- potatoes, sweet potato, beans, peas etc., which contain carbohydrates, are called as energy foods.

Nutraceuticals

• The term Nutraceutical was coined by Dr. Stephen De Felice in 1989.

Nutraceutical

Nutrition → Pharmaceutical

- Nutraceutical is any substance that may be considered as food or part of a food and provides medical or health benefits encompassing prevention and treatment of diseases.
- They are also referred to as a product isolated or purified from foods.
- They are generally sold in forms of medicine and are not usually associated with food.
- They provide a physiological benefit or provide protection against chronic diseases.

Nutrition Enhanced

High yielding & Lycopene rich Tomato variety

Tomato Carotenoids & Antioxidants

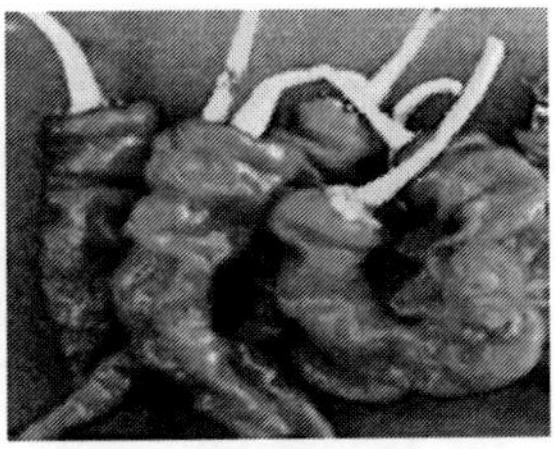

Pepper Carotenoids & Antioxidants

Source of Nutraceuticals

Vegetables	Nutraceuticals
Allium vegetables (Garlic, Onions, Chives and Leeks)	Allyl sulfides
Cruciferous vegetables (Broccoli, Cauliflower, Cabbage, Brussels Sprouts, Kale, Turnips, and Kohlrabi)	Indoles / glucosinolates, Sulforaphane Isothiocyanates / thiocyanates and Thiols
Solanaceous vegetables (Tomatoes and Peppers)	Lycopene
Umbelliferous vegetables (Carrots, Celery, Parsley and Parsnips)	Carotenoid, Phthalides and Polyacetylenes.
Compositae plants (Artichoke)	Silymarin
Beans	Flavonoids (isoflavones)
Carrot, Squash, Broccoli, Sweet Potatos, Tomato , Kale and Pumkin	Vitamin A
Cabbage and Tomato	Vitamin C
Green leafy vegetables	Vitamin E

Edible colours

- Natural pigments are found in living tissue of plants.
- These colours are the chemical compounds produced by several biochemical pathway and gives colours to the food.
- These colours are changed according to the growth stage of the plants or vegetable product.
- These includes anthocyanins, betalains, carotenoids and chlorophylls.

Anthocyanins

- The most extensively documented phytomedicinal role of anthocyanin pigments is in improving eye sight, including night vision.
- Anthocyanins exert significant antimicrobial properties.
- In association with other flavonoids quite effectively inhibits the biosynthesis of aflatoxins.

Betalains

- Betalains have been widely used as natural colourants for many centuries.
- It is having high anti-oxidative properties with free radical scavenging activities.
- It concerns about the use of various synthetic alternatives.

Carotenoids

- Carotenoids provide protection to vision and eye function.
- It acts against macular degeneration and cataracts.
- They are credited with biological promotion of immune system response.
- They are also associated with inhibition of several types of cancers including cervical, oesophageal , pancreatic, lung, prostate, colorectal and stomach.

Chlorophylls

- Chlorophyll is the most important plant pigment and a real life force that nature uses to explode plants into greenery.
- It is consumed much higher doses in diet with green and leafy vegetables.
- The intake of chlorophyll has potential use to act as a chemo preventive compound in humans due to its anti-mutagenic properties.

Vegetables rich in pigments

Colour	Principle Compound	Vegetables
Red	Lycopene	Tomato and Watermelon.
Orange	β-carotene β-cyanins	Carrot, Melon, Red Pepper, yellow and orange-fleshed Sweet Potato. Beet root.
Green	Chlorophyll	Broccoli, Kale, Spinach, Cabbage Asparagus and Green Leafy Vegetables.
Blue/ Purple	Anthocyanins	Red Cabbage, Brinjal, *Basella rubra* Purple Tomato, Purple Sweet Potato and Black Carrot.
Yellow	Lutein	Yellow Corn.
Black / Brown	Anthocyanins	Black Carrots.

Some of the essential nutrients provided by different fruits are

Vitamins Minerals	Role in human body	Source
Vitamin-A	1. Essential for growth and reproduction. 2.Helps in resistance to infections, increases longevity and decreases senility. 3.Deficiency causes, night blindness, xeropthalmia, retardation in growth, roughness in skin, formation of stones in kidney.	*Mango, Papaya, Persimon, Dates, Jackfruit, Walnut, Oranges, Passion fruit, Loquât, Coriander leaves, Drumstick leaves, Fenugreek leaves etc.*
Vitamin – B1	1. For maintaining good appetite and normal digestion. 2. Necessary for growth, fertility, lactation and for normal functioning of nervous system. 3. Deficiency causes beri-beri, paralysis, loss the sensitivity of skin, enlargement of heart, loss of appetite and decline in body temperature.	*Walnut, Apricot, Apple, Banana, Grapefruit, Plum, Almond Chillies, Colocasia leaves, Tomato, etc.*
Vitamin – C	1. Deficiency causes scurvy, pain in joints, swelling of limbs, unhealthy gums, tooth decay, delay in wound healing and rheumatism.	*Barbados cherry, Aonla, Guava, Lime, Lemon, Sweet oranges, Ber, Pineapple, Pea,Chillies, Tomato, Coriander leaves, Drumstick leaves etc.*
Fat	Fat is required for lubrication of joints and maintaining the body temperature	*Walnut, Almond, Avocado.*
Fiber	For proper digestion	*Guava, Pomegranate, Aonla, Grape, Amaranth, Mustard, Beet leaf, Spinach etc.*

Minerals essential for the growth and development for the human body

Minerals	Deficiency causes	Sources
1. Calcium	Causes Rickets, Osteomalacia	*Sitaphal, Ramphal, Fig, Phalsa, Citrus, Sapota, Grapes, West Indian Cherry etc. Curry leaves, Amaranthus, Radish leaves, Fenugreek leaves et.c*
2. Phosphorous	Essential for cell multiplication of bones and soft tissues. Helps in liberation of energy on oxidation of carbohydrates.	*Wood apple, Avocado, Dates, Pomegranate and Grape raisins.*
3. Proteins	Important for body growth, formation and maintenance of body tissues	*West Indian cherry, Avocado, Custrad Apple, Banana, Apricot, Guava, Grapes etc., Peas, Cowpea, Bean etc.*
4. Iron	Act as oxygen carrier in the body.	*Karonda, Date palm, Grape raisins, West Indian Cherry. Guava, Sitaphal, Avocado, Sapota, Plum, Amaranthus, Coriander leaves etc.*

Nutritive value of some important fruits and vegetables

Nutrient	Crop	g Per 100 g
Protein	Agathi	8.4
	Drumstick leaves	6.7
	Field beans	3.8
Fat	Avocado	22.8
	Cabbage	2.1
	Drumstick leaves	1.7
Minerals	Agathi	3.1
	Araikeerai	2.4
	Drumstick leaves	2.3
Fiber	Chillies green	6.8
	Guava	5.2
	Grapes	2.8
Carbohydrate	Dates - fresh	33.8
	Sweet potato	28.2
	Banana	27.2
Energy	Avocado	215 K.cal
	Dates - fresh	144 K.cal
	Sweet potato	120 K.cal

Fruits and vegetables rich in minerals and vitamins

Nutrient	**Crop**	**mg per 100 g**
Calcium	Agathi, Drumstick leaves	1130440
Phosphorous	Carrot, Avocado, Agathi, chillies	53880
Iron	Potato	40.0
	Araikeerai	38.5
Carotene	Drumstick leaves, Agathi	67805400
Thiamine	Potato, Agathi	240.21
Riboflavin	Chillies, green Papaya	0.390
Niacin	Agathi, Potato	1.21.2
Vitamin C	Drumstick leaves, Guava	220212

3

Horticultural Zones

The Horticultural crops show wider adaptation and therefore exact delineation of the agro-climatic areas specific to each crops is not possible, except for the typical temperate and tropical species. However, our country can be broadly divided into seven zones for horticultural crops as indicated below:

1. Temperate Northern Region

Comprising of states of Jammu & Kashmir, Himachal Pradesh, Hilly areas of Uttar Pradesh and West Bengal. Temperate fruits, cool season vegetables are the important horticultural crops in the zone.

2. North Western Arid Region

Comprising of the entire Rajasthan state, Gujarat and parts of Punjab & Haryana. Arid fruit crops like ber, pomegranate, amla, date palm, wood apple and seed spices like coriander, cumin, fennel occupy major area in this zone.

3. North Eastern Sub-tropical Humid Region

Comprising of states and union territories like Arunachal Pradesh, Assam, Manipur, Meghalaya, Mizoram, Nagaland and Tripura. Fruit crops like banana, pineapple, citrus, jack besides plantation crops like tea and large cardamom are also grown here.

4. North Central Sub-tropical Region

Includes parts of Uttar Pradesh, Bihar, entire Madhya Pradesh and parts of Maharashtra. Fruit crops like mango, sapota, sweet oranges and guava are grown here.

5. South Central Tropical Region

Includes the central area bordered by the Western and Eastern Ghats in the states of Tamil Nadu, Andra Pradesh, Karnataka and parts of Maharashtra. Mango, guava, sapota, pineapple and turmeric are chiefly grown here.

6. Coastal Tropical Humid Region

The entire coastal region stretching along the Bay of Bengal in the East and Arabian sea in the West comprises of this region. Banana, mango, cashew, coconut are the major crops.

7. Southern Hilly zone

Comprising of Western ghat and Eastern ghats hills above 800 m MSL, where coffee, tea, cardamom, pepper, oranges and pineapple are extensively grown.

Horticultural zones in Tamil Nadu

Tamil Nadu state is situated at the South-Eastern extremity of the Indian peninsular bounded on the North by Karnataka and Andhra Pradesh, in the east by Bay of Bengal, in the South by Indian Ocean and in the West by Kerala State. It has a coastal line of 1076 km and a land boundary of 1200 km. It lies between 8 5' and 13 35' at Northern Latitude and 76 15' and 80 20' of Eastern Longitude with an area of 1,30,058 sq. km. (50,154.7 sq.miles).

Physical features

The state is divided broadly into two natural division ie., (a) the coastal plains and (b) the hilly western areas. It is further divided into coromandal plains comprising the districts of Kancheepuram, Tiruvellore, Cuddalore, Villupuram and Vellore, Thiruvannamalai and alluvial plains of Cauvery delta extending over Thanjavur, NagapattinamThiruvarur and part of Trichy and dry Southern plains in Madurai, Ramanathapuram, Sivaganaga, Virudunagar, Tuticorin and Tirunelveli districts. It also extends a little in Western Ghats in Kanayakumari district. The Western Ghats averaging 3000' to 8000' height runs along the western part with the hill zone of Nilgiris and Anamalais on either side of it. Palani hills, Varashanad and Andipatti ranges are the major off-shoots of Ghats. The other prominent hills comprise of Javadhu, Shervarayan, Kalrayans and Pachaimalais. These ranges continue South of river Cauvery. A plateau is found between these hills and the Western Ghats with an average elevation of 1000 feet raising West-ward. The highest peak Doddapettah in the Nilgiris is 8650' feet above M.S.L.

Western Ghats forms a complete watershed and no river passes through them. The main streams i.e., Paraliyar, Vattasery Phazhayar etc., are of limited length and fall in the Arabian sea. All other rivers are east-flowing rivers. The Eastern Ghats are not a complete watershed and as a results the river pass through them at places, notable among them is the river Cauvery. The main river of Tamil Nadu are Cauvery (with tributaries of Bhavani, Amaravathi, Noyyal) Vaigai, Tamaraparani, Palar, Ponniyar and Vellar.

Climate

The climate of Tamil Nadu is basically tropical. Due to its proximity to the sea, the summer is less hot and winter is less cold. In plains, the maximum daily temperature is 44.4°C and the minimun daily temperature is 11.8°C. In hill station, maximum temperature is 25.5°C and minimum temperature is 4.4°C. The state is exposed to both South West and North East monsoon.

Agro-climatic Zones

The seven agro climatic zones viz., Western, Southern, North Eastern, North Western, Delta, High Rainfall, Hilly and Tribal zones possess conducive climatic condition coupled with monsoon rains.

Agro-climatic zones of Tamil Nadu

Sl.No.	Agro-Climatic Zone	District
1.	Western zone	1. Coimbatore
		2. Erode
		3. Dindigul
		4. Theni
2.	Southern zone	1. Madurai
		2. Ramanathapuram
		3. Tirunelveli
		4. Sivagangai
		5. Virudhunagar
		6. Thoothukudi
		7. Pudukottai
		8. Karur
3.	North Eastern zone	1. Kancheepuram
		2. Thiruvallur
		3. Vellore
		4. Thiruvannamalai
		5. Cuddalore
		6. Villupuram
4.	North Western zone	1. Dharmapuri
		2. Krishnagiri
		3. Salem
		4. Namakkal
		5. Perambalur
5.	Delta zone	1. Thanjavur
		2. Thiruvarur
		3. Nagapattinam
		4. Tiruchirapalli
6.	High Rainfall zone	1. Kanyakumari
7.	Hilly and Tribal zone	1. Nilgiris

The special feature of climatic condition in Tamil Nadu is that the four seasons are not distinct and overlap with immediate next season. In the plains there is no winter seasons and in the hills summer is not severe allowing continuous production of horticultural crops in the state.

The summer (March-May) overlaps into windy seasons (June-August) which overlaps into rainy seasons (September-November). The rainy season overlaps into winter (December-February) which in turn overlaps into summer season. There are areas with temperate climate (above 2000 m), subtropical climate (1000-2000 m), humid tropical climate 500- 1000 m and tropical climate (up to 500 m) to accommodate all possible types of horticultural crops.

These overlapping seasons have made possible grapes to grow round the year, two to three crops in a year. Because of differential climate, tropical, sub-tropical and temperate crops can be grown. Vegetables and spices can be grown in all possible crop rotation and systems of planting. Since close to the equator and the presence of long stretch of Western Ghats, discontinuous Eastern Ghats and presence of hill and hillocks in the plains help moderation of climate to ensure the required heat unit and degree-days for growing different kinds of horticultural crops. Absence of heavy wind, frost, hailstorm helps in growing crops throughout the year.

Institutions Involved in Horticulture

Realizing the importance of Horticulture, Indian Government encourages a strong institutional support for Research and Development in Horticulture with research infrastructure spread over in:

- 10 central institutes with 24 regional stations.
- 7 crop specific National Research Centers.
- 10 Multi-disciplinary institutes.
- 13 All India Coordinated Research Projects with 215 centres.
- 36 State Agricultural Universities.
- 15 central general/deemed universities with horticulture discipline dealing with research on different horticultural crops.

Besides research, some of these institutes particularly State Agricultural Universties have to fulfill its another mandatory function *viz.,* generating qualified human resources in Horticulture as they offer B.Sc. (Hons.) Hort. degree programmes.

Central Research Institutes

1. Central Institute for Subtropical Horticulture (CISH), Lucknow, Uttar Pradesh.

2. Central Institute for Temperate Horticulture (CITH), Srinagar, Jammu & Kashmir.
3. Central Institute for Arid Horticulture (CIAH), Bikaner, Rajasthan.
4. Central Tuber Crop Research Institute (CTCRI), Thiruvananthapuram, Kerala.
5. Central Plantation Crop Research Institute (CPCRI), Kasargode, Kerala.
6. Central Potato Research Institute (CPRI), Shimla, Himachal Pradesh.
7. Central Research Institute for Chikoo (CRIC), Muzzaffarpur, Bihar.
8. Central Coffee Research Institute (CCRI), Chikmangalur, Karnataka.
9. Central Arid Zone Research Institute(CAZRI), Jodpur, Rajasthan.
10. Central Research Institute for Post-Harvest Engineering and Technology (CIPHET) Ludhiana.

National Research Centers (NRC)

1. Banana – Tiruchirapalli, Tamil Nadu.
2. Citrus – Nagpur, Maharastra.
3. Grapes – Pune, Maharastra.
4. Litchi – Muzaffaarpur, Bihar.
5. Pomegranate – Solapur, Maharastra.
6. Seed Spices – Ajmer, Rajasthan.
7. Orchids – Pakyong, Sikkim.

Multi-disciplinary institutes

1. Indian Institute of Horticulture Research (IIHR), Bengaluru, Karnataka
2. Indian Agricultural Research Institute (IARI), Pusa, New Delhi.
3. Indian Institute of Spice Research (IISR), Calicut, Kerala.
4. Indian Institute of Vegetable Research (IIVR), Varanasi, Utter Pradesh.
5. Directorate of Cashew Research – Puttur, Karnataka.
6. Directorate of Onion and Garlic Research –Pune, Maharashtra.
7. Directorate of Oil Palm Research- Pedavegi, West Godawari.
8. Directorate of Medicinal and Aromatic Plants Research – Anand,Gujarat.
9. Directorate of Floriculture Research – Pusa, New Delhi.
10. Directorate of Mushroom Research – Solan, Himachal Pradesh.

All India Coordinated Research Projects (AICRP)

1. Fruits – Bengaluru
2. Arid zone Fruits – Bikaner
3. Mushroom – Solan
4. Vegetable Including NSP of Vegetables – Varanasi
5. Potato – Shimla
6. Tuber Crops – Thiruvananthapuram.
7. Palms – Kasargode
8. Cashew – Puttur
9. Spices – Calicut
10. Medicinal and Aromatic Plants including Betal Vine- Anand
11. Floriculture – New Delhi
12. Post-Harvest Technology – Ludhiana.

Horticultural Institutions in Tamil Nadu

1. Horticultural College & Research Institute, (TNAU), Coimbatore
2. Horticultural College & Research Institute, (TNAU), Periyakulam
3. Horticultural College & Research Institute for Women (TNAU), Tiruchirapalli.

Horticulture Research Stations in Tamil Nadu

1. Horticultural Research Station (TNAU), Ooty.
2. Horticultural Research Station (TNAU), Kodaikanal.
3. Horticultural Research Station (TNAU), Thadiyankudisai.
4. Horticultural Research Station (TNAU), Yercaud.
5. Horticultural Research Station (TNAU), Pechiparai.
6. Flower Research Station (TNAU), Thovalai.
7. Vegetable Research Station (TNAU), Palur.
8. Coconut Research Station (TNAU), Vepankulam.
9. Coconut Research Station (TNAU), Aliyar.

4

Commodity Boards and Institutes

The Government of India has established boards for important crops. The boards are statutory bodies which help in guiding the production and export. The main objectives for setting up the boards are –

- To monitor and promote the export of primary items.
- To empower the producers by providing trainings and close guidance.
- To conduct research to improve the commodity, e.g. Central Coffee Research Institute, Coir Research Institute etc.
- To advise the government on policy matters such as price fixing, fixing quota for export and signing trade agreements.
- To promote the commodity by opening offices and outlets in India and abroad.
- To showcase the commodity by participating in trade fairs and exhibitions.

Coffee Board, Government of India, Ministry of Commerce and Industry

- During 1940's, the coffee industry in India was in a desperate state due to the II World war, resulting in very low prices and ravages of pests and diseases.
- At this time, the Government of India established the 'Coffee Board' through a constitutional act "Coffee Act VII of 1942" under the administrative control of Ministry of Commerce and Industry.
- The Board was entrusted with the responsibilities of extending support to the coffee industry in the areas of marketing, finance, research and development.
- Accordingly, the Coffee Board took over the control of the 'Mysore Coffee Experiment Station' in the year 1946 and renamed it as Central Coffee Research Institute.
- This Institute was made as the Head Quarters of the Research Department of the Coffee Board with a larger mandate of undertaking extensive research on coffee covering different disciplines and dissemination of technology to the growers from time to time.

Tea Board, Government of India, Ministry of Commerce and Industry

- As per the provision of Tea Act (1953), Tea Board of India supports and promotes tea research for the development of Indian tea industry.
- The Research Directorate of Tea Board has been entrusted to conduct, coordinate and evaluate tea research nationally through three tea research institutes namely -
  - Darjeeling Tea Research and Development Centre (DTR&DC), Kurseong, Darjeeling, West Bengal.
  - Tea Research Association (TRA), Jorhat, Assam.
  - United Planters Association of Southern India - Tea Research Foundation (UPASI-TRF).
- The DTR&DC is engaged in the Research and Development (RaD) activities to cater to the needs of the tea industry of Darjeeling.
- TRA and UPASI are involved to look after the requirement of North East India and South India respectively.
- TRA and UPASI, each having seven advisory centres scattered throughout different tea plantation areas in their respective zones.
- Apart from the above mentioned three tea research institutes, research projects are being given to other national R&D Institutes and Universities to conduct research for the benefit of Indian tea industry.
- Tea research in India covers basic, applied and regulatory research aspects such as -
  - Plant improvement (breeding and biotechnology).
  - Plant production (agronomy and soil science), Plant protection (mycology and entomology)
  - Tea quality covering biochemistry, tea tasting and tea processing, electronics and engineering.

The vision of the Directorate is to produce quality tea on sustainable basis involving low inputs and state-of-the-art technology through value research.

National Horticulture board

- National Horticulture Board (NHB) was set up by Government of India in April 1984 on the basis of recommendations of the "Group on Perishable Agricultural Commodities", headed by Dr M. S. Swaminathan, the then Member (Agriculture), Planning Commission, Government of India.
- The NHB is registered as a Society under the Societies Registration Act 1860, with its headquarters at Gurgaon.

- The main objectives of the NHB are to improve integrated development of Horticulture industry and to help in coordinating, sustaining the production and processing of fruits and vegetables.

Objectives

- Development of hi-tech commercial horticulture in identified belts and make such areas vibrant with horticultural activity, which in turn will act as hubs for development of horticulture.
- Development of modern post-harvest management infrastructure as an integral part of area expansion projects or as common facility for cluster of projects.
- Development of integrated, energy efficient cold chain infrastructure for fresh horticulture produces.
- Popularization of identified new technologies/tools/techniques for commercialization/adoption, after carrying out technology and need assessment.
- Assistance in securing availability of quality planting material by promoting setting up of scion and root stock banks/mother plant nurseries and carrying out accreditation/rating of horticulture nurseries and need based imports of planting material.
- Promotion and market development of fresh horticulture produce.
- Promotion of field trials of newly developed/imported planting materials and other farm inputs; production technology; PHM protocols, INM and IPM protocols and promotion of applied R&D programs for commercialization of proven technology.
- Promotion of Farm Mechanization in Horticulture through demonstration and its uses at farmer's field level to reduce labour cost and increase the productivity of Horticulture crops.
- Promotion of applied R&D for standardizing post harvest management protocols, prescribing critical storage conditions for fresh horticulture produce, bench marking of technical standards for cold chain infrastructure etc.
- Transfer of technology to producers/farmers and service providers such as gardeners, nurserymen, farm level skilled workers, operators in cold storages, work force carrying out post-harvest management including processing of fresh horticulture produce and to the master trainers.
- Promotion of consumption of horticulture produce and products.

Spices board

- Spices board was constituted on 26th February, 1987, under the Spices Board Act 1986, with the merger of Cardamom board (1968) and Spices Export Promotion Council (1960).
- It is responsible for promotion and export of 52 spices and development of small and large Cardamom.

It has the following mandates –

- Research, development and marketing of small and large cardamom.
- Promotion of organic spice production and export.
- Improvement in postharvest management and processing.
- Organic and quality certification of spices.
- Quality improvement of spices.
- Technology upgradation.
- Promotion and advertisement of spices in India and abroad.
- Guidance to farmers.
- Financial, material and technological support to the growers.
- Facilitating infrastructure for processing and value addition.
- Registration and licencing of exporters.
- Exporter and importer interface
- Mediating and facilitating exports.
- Bridging the gap between international bodies, exporters and policy makers.
- Participation and advertisement in international arena through trade fairs.

Coir Board of India

- The board is a statutory body established by Government of India under the Coir Industry Act, 1953.
- It functions under Ministry of Micro, Small and Medium enterprises.
- The head office is in Kochi.
- The research and training office is at Kalavoor, Alapuzza.
- It undertakes promotion and development of coir industry.

Rubber board

- This board was constituted by the government of India under the Rubber Act of 1947. The head office is in Kottayam, Kerala.
- It under takes planning, acquisition and marketing of Rubber.

- The board advices the Government in all matters related to rubber industry
- It is responsible for registration of rubber estates, issue of licence for new planting and replanting.
- It also looks after the subsidy schemes.

National medicinal plant board

- NMPH was established on 24thNovember, 2000. It is located in Ministry of Ayush, Government of India.
- It is advisory body which helps in strategic planning of medicinal plant related initiatives
- It plans and supports programmes related to conservation and development of medicinal plants

Role of APEDA in Horticulture commodities

- Agricultural & Processed Food Products and Export Development Authority (APEDA) is with Ministry of Commerce and Industry, Govt. of India, New Delhi.
- APEDA was established by the Government of India under the Agricultural and Processed Food Products Export Development Authority Act passed by the Parliament in December, 1985.
- The Act (2 of 1986) came into effect from 13th February, 1986 by a notification issued in the Gazette of India: Extraordinary: Part-II [Sec. 3(ii): 13.2.1986).
- The Authority replaced the Processed Food Export Promotion Council (PFEPC).
- The mandated responsibility of APEDA is export promotion and development of fruits, vegetables, floriculture, herbal, medicinal products and their value added products along with many other consumables with export potential

Functions

- Development of industries relating to the horticultural products for export by way of providing financial assistance
- Undertaking surveys and feasibility studies
- Participation in enquiry capital through joint ventures
- Other reliefs and subsidy schemes
- Registration of persons as exporters of the horticultural products on payment

of such fees as may be prescribed

- Fixing of standards and specifications for the horticultural products for the purpose of exports
- Carrying out inspection of horticultural products, processing plants, storage premises, conveyances or other places where such products are kept or handled for the purpose of ensuring the quality of such products
- Improving of packaging of the horticultural products
- Improving of marketing of the scheduled products outside India
- Promotion of export oriented production and development of the horticultural products
- Collection of statistics from the owners of factories or establishments engaged in the production, processing, packaging, marketing or export of the scheduled products or from such other persons as may be prescribed on any matter relating to the scheduled products and publication of the statistics so collected or of any portions thereof or extracts therefrom
- Training in various aspects of the industries connected with the horticultural products and such other matters as may be prescribed.

5

Garden Tools and Implements

A number of tools, implements and machineries are used in cultivation of horticultural crops to make every operation easy and efficient. There are a number of operations involved in cultivation of horticultural crops, like making of nursery, propagation of plant material, field ploughing, breaking of clods, levelling, making bunds, sowing and planting, plant protection, training, pruning, harvesting, lawn mowing etc. All these operations are physically taxing and cannot be carried out efficiently without proper tools, implements and machineries. Following are some important tools -

1. **Bill hook**: To cut and remove big sized unwanted branches of trees.
2. **Buckets and plastic pots**: used in the orchard for spraying, transport water, carry harvested produce etc.
3. **Budding and grafting knife**: The budding and grafting operations in propagation of many horticultural plants are carried out using budding and grafting knife.
4. **Country plough**: Used in field preparation for ploughing the soil, incorporation of organic manures into the soil at the time of last ploughing, summer ploughing in conserving of soil moisture.
5. **Crow bar**: Used to loosen the soil, digging and deepening the pits and channels.
6. **Dibbler**: Used to make holes and sow the seeds in the nursery as well as in the main field.
7. **Digging fork**: It is of a great use in harvesting the tuber crops without damaging the produce.
8. **Fruit harvester**: The fruits from tall tree branches are harvested with fruit harvester. The fruits are plucked with the help of a sharp knife attached at the tip of the long handle. The fruit is collected into a nylon net bag attached to it. Fruit harvester helps in harvesting the fruits without damage (or) bruise. Specialized fruit harvesters are also available for specific crops such as like coconut harvester for coconut. It is a mechanized equipment which helps the harvester to reach up the tall tree and harvest coconuts.

9. **Garden fork**: This is also used in nursery beds to remove deep rooted weeds and to loosen the soil.
10. **Garden rake**: Garden rake is mainly used in the nursery to rake up the soil and clean surface debris.
11. **Grass shear**: This helps cut the overgrown grass, when the lawn mower is not advisable to use.
12. **Hand hoe**: Mainly used for weeding and loosening soil in the main field.
13. **Hedge shear (or) garden shear**: The height and shape of the hedge in a garden is maintained by pruning the excess growth on all sides using hedge shear. Also used to remove unwanted shoots of horticultural crop plants.
14. **Iron pan**: Iron pans are used to transport soil, manures and fertilizers etc.
15. **Lawn mowers**:Used to mow the lawn grass to keep grass at desired manageable height. Lawn mowers can also be used to adjust the cutting height of the grass and mow the lawn grass in different heights, to give it light and dark shades,in stripes, circles and other patterns to add beauty to the lawn.
16. **Mud pots**: Mud pots of different sizes are used to plant young plants, seedlings, cuttings, raising rootstocks etc. Seedlings of different crops are also raised in mud pots especially seed pans and after specified period they are transplanted in the main field. The ornamental plants potted in mud pots are displayed to add beauty to the surroundings.
17. **Pitchers and pots**: Used to water the propagated plants in the main field and carry water for spraying.
18. **Pruning knife**: The shoots of young plants and shrubs are removed (or) pruned with the help of the pruning knife.
19. **Pruning saw**: It is used to cut very big branches with a clean cut without any scrapping.
20. **Pruning shear**: Used to prune (or) remove the shoots which are of 5-6 cm diameter in size.
21. **Ropes**: In the orchard, ropes are used to lay out the field, to mark lines for sowing, planting and various operations in straight rows.
22. **Rose can**: Watering of young plants kept in the nursery. Fine spray of water comes out through the rose attached with the can.
23. **Scythe**: The tall weeds in the perennial plots, along the roadsides are removed using scythe in the orchard.
24. **Secateur**: Used in vegetative propagation of many horticultural crop species. Also used in training and pruning operations to remove unwanted shoots, dried, pest and disease affected twigs of a plant.

25. **Seed trays**: The extracted seeds of all horticultural crops are dried after spreading on the seed trays. Trays of different sizes are available for this purpose.
26. **Shovel**: Used in the nursery to transport soil.
27. **Sickle**: Used in harvesting the produce of many horticultural crops. The tall grown grasses also removed using sickle.
28. **Spade**: Used in many field and nursery operations. To dig pits, form ridges and furrows, irrigation channels, drainage channels, nursery beds, to loosen soil and to scrape off weeds.
29. **Sprayers**: Sprayers are of great use in the cultivation of horticultural crops. They are used to spray pesticides, fungicides, herbicides, growth regulators, etc. to crops in main field, tree species, plants in the nursery, home garden, indoor plants etc. The commonly used sprayer in the orchard are i) Ganesh sprayer, ii) Rocker sprayer, iii) Knapsack sprayer, iv) Power sprayer and v) Tree sprayer.
30. **Stakes**: Used for lay out of the field, to mark pits and for labeling the plants.
31. **Tractor trolley mounted hydraulic platform:** This hydraulic machine can be used to move inside the orchard and reach to a height of 25 feet from the ground. Pruning can be done using pruning saw, tree pruner and chain saw by standing in the platform provided.
32. **Transplanting trawl**: Used in transplanting the seedlings.
33. **Tree pruner**: The small branches of trees of 10 cm (or) more in diameter are pruned with tree pruner.
34. **Wheel barrow**: Wheel barrow is used in many ways in the orchard. It is used to transport pot mixture, manures and fertilizers, pots, harvested produces, plants etc.

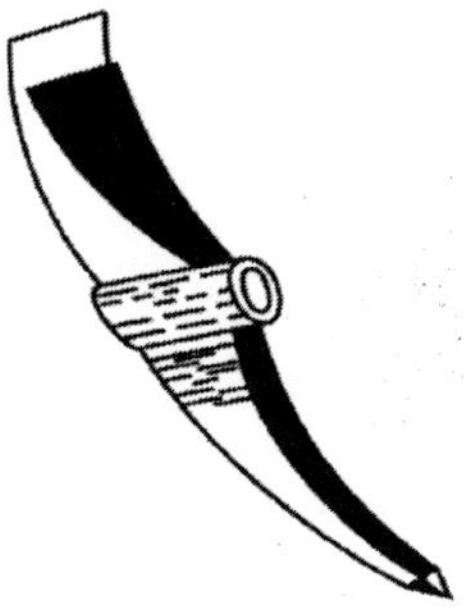

Pick Axe

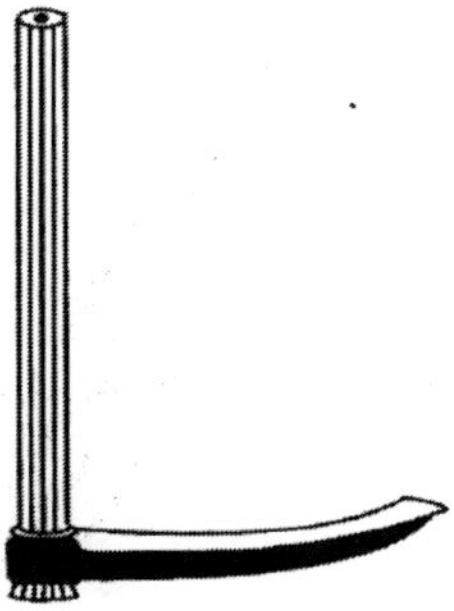

Kodali

Fork

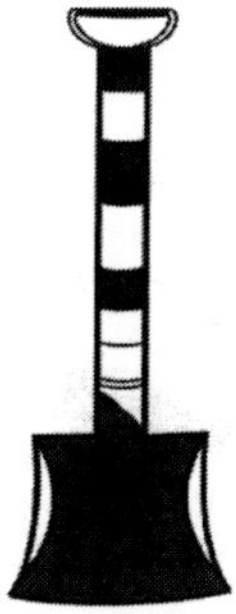

Shovel

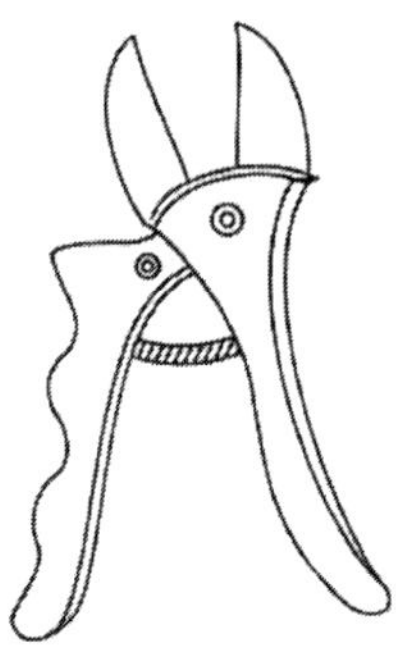
Secateur

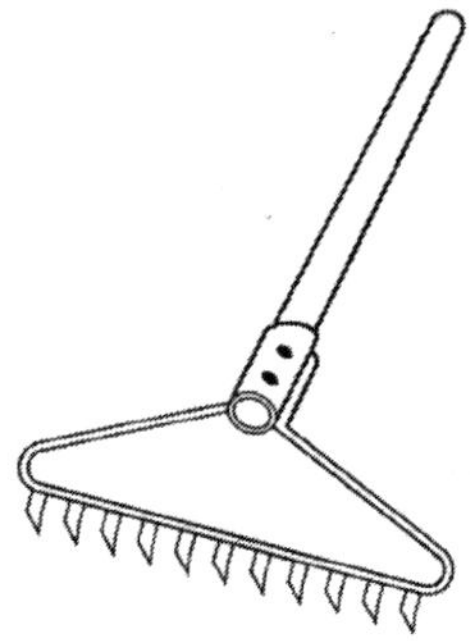
Garden Rake

Flower Scissor

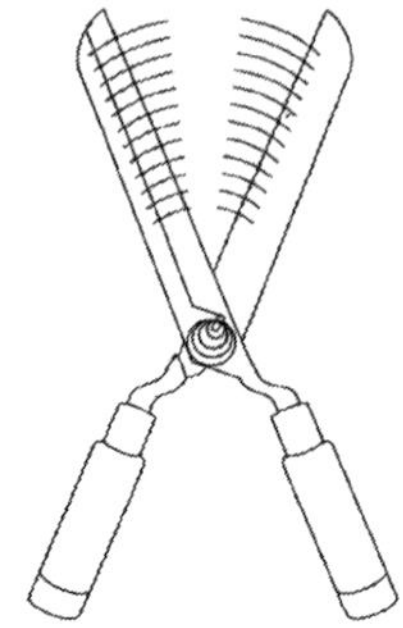
Leafy Vegetable Harvester

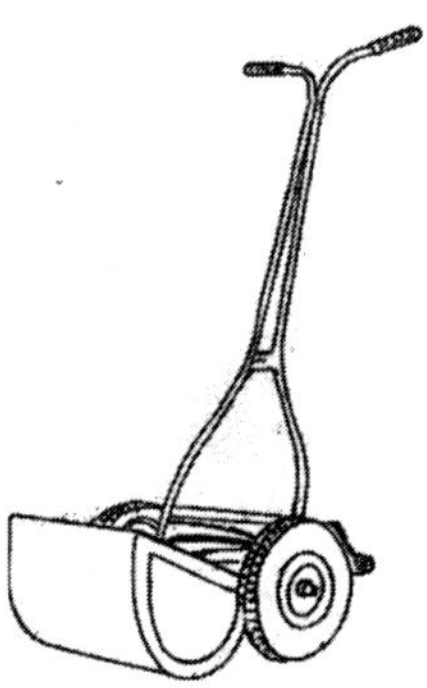
Lawn Mower

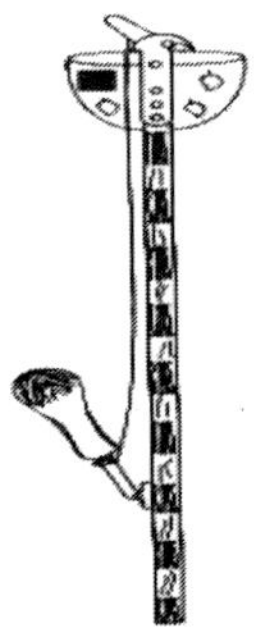
Mango Harvester

Hand Sprayer

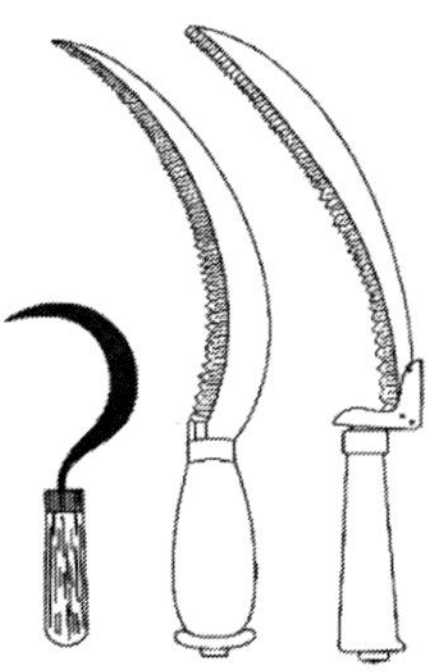
Sickle

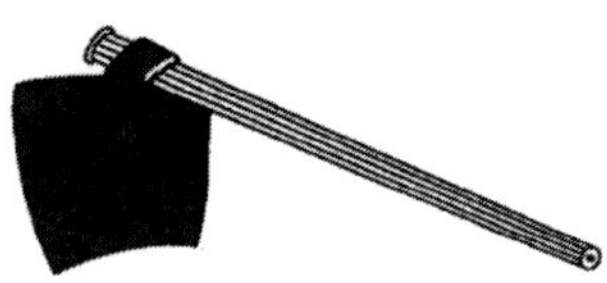
Spade

Pruning Knife

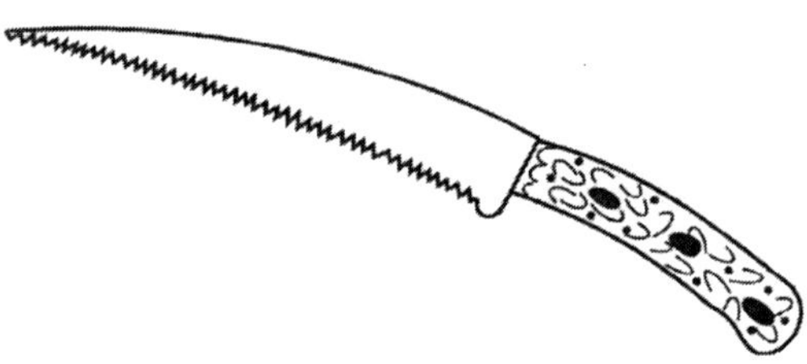
Pruning Saw

Hand leveler

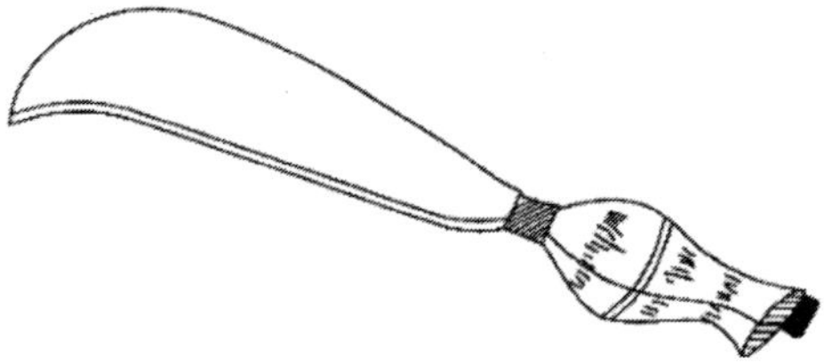

Bill Hook

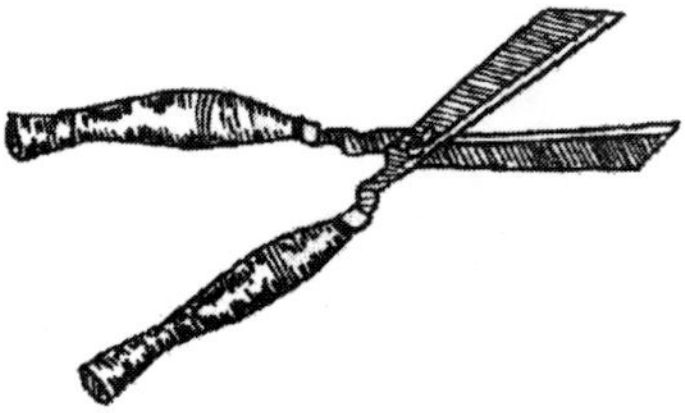

Hedge Shear

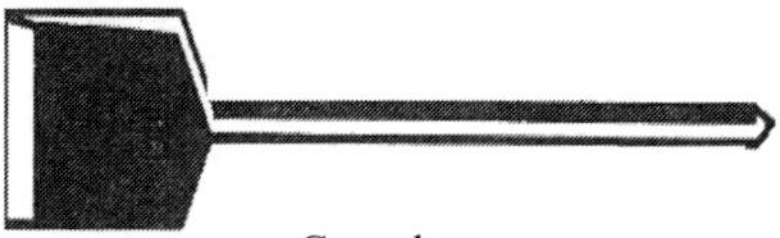

Crow bar

Axe

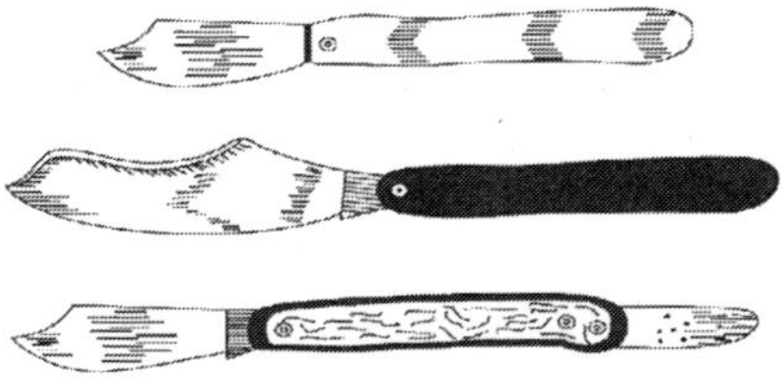

Budding Knife

Budding and Grafting Knife

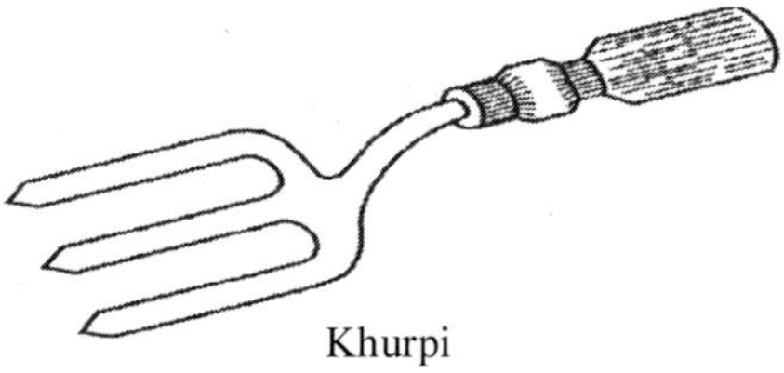

Khurpi

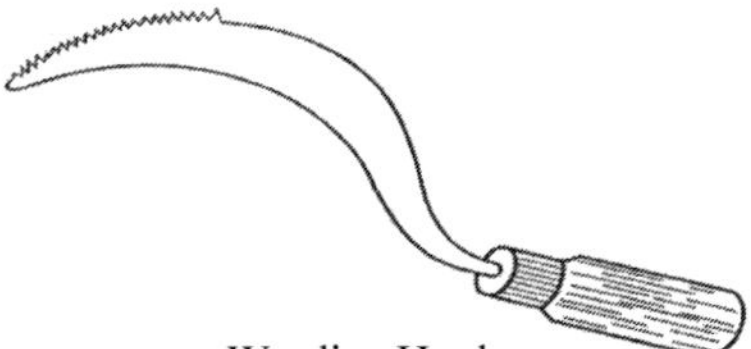

Weeding Hook

Foot Sprayer

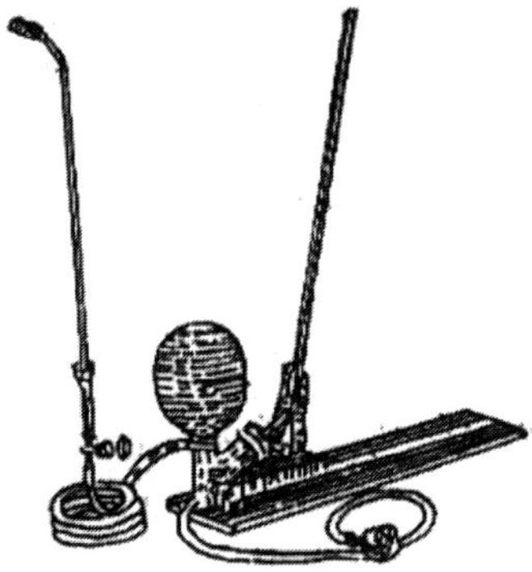

Rocker Sprayer

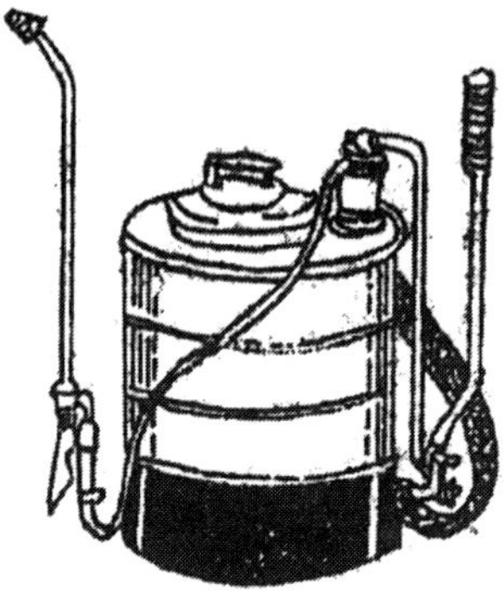

Knap – Sack Sprayer

Grafting Knife

6

Impact of Soil and Environment on Productivity

The performance of plants is affected by the soil in which it grows and climate and micro climate of its surroundings.

Impact of soil

The soil provides anchorage, nutrients and water for plant growth. Soil fertility status, structure and texture influence the plant growth and productivity.

Classification of soil

On the basis of the organic and inorganic contents and particle size and texture, the soils have been classified. The inorganic matter in the soil constitutes of varying amount of sand, silt and clay, on the basis of which the soils are classified as follows.

Sandy soil

These soils have coarse textured and large pore spaces, have poor water holding capacity. Retention of nutrients in this soil is very poor. This type of soil is suitable for precocious crops. Sandy soils initiate rapid root growth, hence good as propagating medium. Since sandy soils has low in nutrient content, supplementation of adequate nutrition is imperative to achieve good plant growth.

Loamy soils

Loamy soils composed of sand, shilt and clay. Based on the percentage of sand, silt and clay in the loamy soils, they are categorized as sandy loam, silty loam or clay loam. Loamy soils are best for plant production.

Clay soils

These are fine textured soils which are compact and have very small pore spaces. Clay soils have high water retention capacity and are rich in nutrients. Because of the compactness, growing root and tuber crops is a constraint in clay soils. When fruit crops like mangoes are grown in rich clay soils, occurrence of heavy vegetative growth is noticed with low fruit production.

Organic matter

It is an important component in soil. It is derived from decomposed plant and animal tissues. Microorganisms act on the plant and animal debris or tissues by enzymatically digesting them and thus decomposition takes place which enriches the soil by supplementing the soil with enough organic matter. The decompositions of lignins, waxes, fats and some proteinaceous materials are difficult, but possible with the help of humus, which is a non-crystalline colloidal substance through complex bio-chemical processes. Humus is rich in nutrients with high water holding capacity.

The organic matter in the soil can be increased by supplementing the soil with decomposable plant debris and also by growing green manure crops. Organic soils are high in organic matter, usually 20% or more. They are found in swamps, bogs and shallow lake bottom and river beds. The organic soils are classified as peats and mucks. The peat soil has 50-95% organic matter and high water holding capacity, while the muck contains 20-50% organic matter and low water holding capacity. In peat soil, the decomposition is faster due to availability of high moisture. Generally the hill soils have more organic matter because there is lot of leaf litter and owing to the terrain and continuous availability of moisture the microbial decomposition of organic matter is faster and undisturbed. Some root, tuber and cole crops like, onion, potato, carrot, radish, beet root, cabbage, cauliflower and lettuce prefer peat soils.

Humic-acid

In recent days, the humic acid, a by product of coal mines extracted from lignite in the form of potassium humate is used as a soil french. This helps in chelating toxic metals, unlock nutrients and bind to them, making then available for uptake into plants. Humic acid improve the structure of soil. It promotes the conversion of nutrient elements into available forms to plants stimulates root growth, enhance better uptake of nutrients.

Soil fertility

It refers the ability of the soil to nourish the plant. The mineral nutrients of the soil become available to the plants when they interact with air, moisture, microbes and humus. The top layer of the earth is the most fertile, because more plant and animal debris are available. Thus, microbial decomposition actively happens in this layer. For the same reason, fertility status of the soil declines in the deeper layers of the soil profile. The mineral nutrients released from the decomposed debris in combination with microbial activity and water decides the fertility status of the soil.

In nature, the fertility of the soil depletes when the crop uses the nutrients for its growth, i.e, crop removal, erosion, leaching and volatilization. Continous cropping generally deplete the soil nutrients. If the same crop is grown continuously, only one kind of nutrient will be depleted from the soil and other unused nutrients may accumulate the level of toxicity. Hence, to keep the soil with balanced nutrition, different kinds of crops needs to be grown.

Erosion, the washing off or blowing away of soil results in loss of large quantities of nutrients from the top soil. Leaching of nutrients down through the soil profile by percolating water on its way to the water table will reduce the fertility level. Volatilization of soil nutrients is the most common to occur. Volatilization of nitrogen, in the ammonia form and also by denitrification process is evident. The soil fertility can be maintained in a good state by following judicious crop and soil management practices. For good performance of crops, soils with good fertility status are preferred.

Chemical Composition of Soil

Soil pH

The acid, alkaline or neutral reaction of a soil is expressed on the pH scale. A pH of 7 is neutral. A pH below 7 is acidic and those above 7 are alkaline. The soil reaction influences the nutrient availability and absorption. For instance, boron is deficient in alkaline soils, but unavailable in very acidic acids. The soil pH also influences the activity of soil bacteria, which in turn affects the plant nutrient availability. Most horticultural crops are adapted to a pH range just below neutral, that is, of a slightly acidic nature.

Optimum pH range for Certain Horticultural Crops

Crops	Optimum pH range
Avocado	5.0-7.0
Banana	5.5-8.0
Citrus	5.5-7.5
Mango	6.0 - 7.5
Beet root	6.0-7.5
Cabbage	5.8-7.3
Carrot	5.8-7.0
Onion	6.2-7.4
Sweet potato	5.2-6.0
Chrysanthemum	6.0-6.5
Rose	6.0-6.5
Tea	4.5-5.0

Certain diseases are promoted in acidic soils while some are promoted in alkaline soils. Club root disease of crops is more prevalent in acidic soils. The alkaline soil pH 7 can be reclaimed by adding aluminiumsulphate, sulphur and calcium sulphate (gypsum), while acidic soils (pH below 7) can be reclaimed by addition by applying hydrated lime or lime stone or magnesium sulphate (epsom). In areas of heavy rainfall, leaching will remove many soluble carbonates resulting in more acid soil.

In alkaline soils concentration of sodium salts (above 0.1%) is dangerous. Information on soil salinity is necessary to select tolerant crop varieties and to adopt proper soil management practices. Relative tolerance of fruit trees to salinity is as below.

Relative tolerance	Level of salinity	Fruit Crops
Tolerant	8 m mhos/cm	Date, Guava, Fig and Grapes
Moderately tolerant	3 to 6 m mhos/cm	Pomegranate, Grape fruit, Apple, Pear and Plums
Sensitive	1.5 to 3 m mhos/cm	Orange, Peach, Avocado, Strawberry

Soil depth

The soil should be at least 2.0 m deep, since roots of most of the fruit trees grow up to 2m deep. In shallow soils, the roots cannot grow very deep and hence, prone to uprooting by heavy winds.

Drainage

Soil drainage depends upon the nature of sub soil. Drainage is an important factor since improper drainage and water stagnation can cause many collar and root infections.

Water table

Water table should be below 2.0 m at all times of the year. A higher water table results in poor aeration in sub soils. Rotting of roots occur due to prolonged submergence of roots in moisture.

Soil microbial properties

Soil is enriched with large number of microbes like algae, fungi and bacteria. Microbes help in decomposing the plant and animal debris and make it as manure by enzymatic activity.

Algae are simple microscopic plants with chlorophyll, living in the soil which trap the energy of sunlight and leave their bodies when dead. These dead bodies accumulate as humus in the soil.

Fungi are minute plants without chlorophyll living as parasites on other plants or thriving on dead matter saprophytically, thus helping further break down of organic material by bacteria.

Bacteria are unicellular organisms less than 1/10,000 of a centimeter in diameter. They multiply at an exponential rate. They are the most useful soil organisms. There are hundreds of species of bacteria.

On the basis of their nature of activity, microbes are classified as follows

- Those which digest the organic matter and generate heat and release ammonia
- Those that convert ammonia into nitrous acid and nitrites by uniting it with the basic elements in the soil.
- Those that convert nitrites to nitrates, in the form that the plants get their nitrogen from the soil through their roots in a state of solution in water.
- Those which are found in the nodules of the leguminous plants. They fix nitrogen directly from air and make it available to the plants.

Microbes are isolated from soil, cultured and multiplied and are being sold as powders in charcoal base as well as liquids. These microbial liquid and powders can be applied to the soil for the assimilation of mineral nutrients. E.g) Azotobacter, *Azospirillum* for nitrogen, Phosphate solubilizing bacteria (PSB) for phosphorous and Potassium solubilizing bacteria (KSB) and *Arbuscular* mycorrhizae fungi (AM fungi). Incorporation of these microbial fertilizers is very important because the soil has become devoid of these beneficial microbes due to continuous use of inorganic fertilizers.

Environmental Factors

Climate and micro climate of the crops influence the performance. Climate includes several basic environmental components such as temperature, rainfall, atmospheric humidity, wind, hail storms etc.

Temperature

Temperature is an important environmental factor that influences photosynthesis, respiration, water absorption, transpiration and cell differentiation and all other life functions of the plant. Every stage in the life of a plant is influenced by temperature. For e.g) flowering, fruiting, fruit ripening, fruit quality is all temperature specific. Temperature higher than required by the plant causes scalding in fruits. Temperature less than the requirement causes chilling injury. In low temperature, the conversion of carbohydrate to sugar is slow and reduces the taste, while fruits harvested during warmer months and sunny days tastes good. Soil temperature above 32°C inhibits root growth and development. Soil temperature can be measured by soil thermometer.

Light

Light is necessary for good fruit colour. High light intensity causes sun burn or sunscald. E.g. greening in citrus in temperate regions and breaking of pseudostem in banana. The duration of light influences many processes in plant growth and development like germination, flowering, sex expression and colour development in fruits.

In cucurbits, long day conditions cause initiation of male flowers, while short day conditions favour production of female flowers. Tomato, when grown in open field develops orange to red colour, while the same variety when grown under greenhouse conditions develop deep red colour under filtered sunlight.

Relative humidity

Based on the relative humidity in the atmosphere, the horticultural crops can be classified as humid, semi-arid and arid-zone crops. Relative humidity has a great role in flowering of plants and sex expression which was observed in papaya. High humidity immediately after rainfall increases the incidence of leaf spot diseases in banana. High humidity aggravates the incidence of pests and diseases in many fruit crops. E.g. Downy mildew in grapes. High humidity with rainfall causes the splitting of fruits in grapes and pomegranate. Maintaining high relative humidity in mist chamber encourages rooting of cuttings in fruit crops.

Rainfall

More than the quantum of rainfall, the distribution of rainfall is important for growth and development. Heavy rains in short term cause complete washing of pollen, hinders the insect movement and thereby pollination. It injures the pollen and dilutes the stigmatic fluid and increases the flower drop. Heavy downpours in a shorter period may also cause pest outbreak. On the other hand, low rainfall causes stunted growth, poor yield, damages flowering and fruitset in many crops. But summer showers in coffee are required because it causes bud break and facilitate gregarious flowering.

Wind

Slight winds in open field are necessary for anther dehiscence and transfer of pollen to stigmatic surface in horticultural crops for increased fruit set. Heavy winds, however, damages the crop physically by uprooting the trees, breaking of branches shedding of flowers and fruits. E.g. Wind speed more than 50 km/

hour (Cyclone) completely damages banana crop. Frequent dry and high winds cause rapid loss of moisture and necessitate frequent irrigation.

Hail storm: Rain along with wind and ice formation in temperate regions cause shedding of flowers and fruits. Example, the hail storms during pre-blooming or blooming period of apple, plum, peaches affect the fruit set severely.

Air pollutants in atmosphere

In recent days, accumulation of air polluting gases in the atmosphere has become a great problem to horticultural crops. The harmful gases such as sulphur-di-oxide, carbon monoxide, chloroflurocarbon from green houses and automobile industries deplete the ozone layer there by causing "greenhouse effect' and raise in global temperature. The air pollutants reduce the assimilation rate, growth and development of plants leading to reduction in yield and quality of the produce.The gases like carbon-di-oxide, sulphur dioxide and acetylene released from brick-kilns are causing black tip disorder in mangoes, reported from Punjab. Uttar Pradesh, Bihar and West Bengal.

Frost

Frost occurs in hill stations situated above 2000m MSL. During frost, a thin layer of ice crystals are deposited on leaves, plants and soil and arrests the photosynthetic activity of the plant, besides causing freezing injury. Frost if occurs especially on clear sunny days, more particularly in orchards located either at the bottom of the slope or in the valley areas near water sources or damp places, the damage to the crop is found to be heavy. In Nilgiris and Kodaikanal, frost normally damages crops like tea, potato, cole crops during winter season.

7

Nutrient Management

Nutrient management is one of the key aspects that ensure good quality produce in optimum quantity.Nutrient availability influences growth and productivity of crops. Sixteen elements are recognized as essential nutrients for growth and development of plants. These essential elements are carbon, hydrogen, oxygen, nitrogen, phosphorus, potassium, calcium, magnesium, sulphur, iron, zinc, manganese, copper, boron, molybdenum and chlorine. While carbon, hydrogen and oxygen are derived from air, soil and water, other elements are taken by the plant roots from the soil, according to the requirement of the plant.The soil contains these elements in varying quantities depending on the origin of the soil. According to the quantity of requirement of these elements by plants for its various growth processes,the elements are classified into different categories.

- Major nutrients or primary nutrients - These are required by plants in large quantities because they are involved in all the processes of plant growth and development.Major nutrients are nitrogen, phosphorus and potassium.
- Secondary nutrients - These are required by plants in relatively smaller quantities for growth and development. Secondary nutrients are magnesium and sulphur.
- Micro nutrients - As the name suggests, these are required by plants in minor quantities for their physiological processes. These elements are iron, zinc, manganese, copper, boron, molybdenum and chlorine.

Each of the plant nutrients has a definite quantitive role and specific function in the physiological processes of plants. If any of the essential element is deficient or absent in the soil sysem, the plant growth is affected negatively and this is reflected as inferior quality produce. On the contrary, if a nutrient is excess in the soil system, it hinders in absorption of other nutrients and also may itself be toxic. Hence to get the required growth, quality and quantity of the produce, optimum availability of all the nutrients should be ensured.

The nutrition requirementof annuals is different from that of perennials. For annuals the need is assessed from seed to seed cycle, while for perennials it is assessed on the basis of nutrient required to produce fruit in a year and again for sustaining the next year's growth. The soil may have nutrients , but whether all essential nutrients are available for plant at a given point of time is important to know. If a soil is deficient in a particular nutrient, it has to be supplemented to the soil as manures and fertilizers and this supply has to be regulated from time to time. Hence, ascertaining the mineral nutrition profile of soil is imperative. Many easy diagnostic techniques are designed to get the complete soil profile, so that one can have knowledge about the deficient and toxic elements and take adequate measures.

Few popular diagnostic techniques are listed below:

- Soil analysis
- Leaf analysis
- Nutrient uptake studies (crop removal)
- Experiments under controlled conditions and field trials

When leaf analysis is done, the leaf sample should be collected before the application of manures and fertilizers. The stage of leaf and the time of leaf collection are different for different crops. Some examples are given below.

Crop	Plant part	Age, stage and position and other
Mango	Leaf with petioles	4-7 months old leaves from middle of the shoot
Banana	Leaf lamina on both sides of midrib from the centre of leaf lamina	3rd fully opened leaf which is around 20 days old
Citrus	leaf	4 months old

Based on the results of the diagnostic techniques, yield and fruit quality, the nutrient status of the soil can be classified into five categories viz., deficient, low, optimum, high and excessive or toxic.

Factors affecting nutrient availability

- **Soil** - soil type, pH, soil microbial status
- Crop - age of the plant, variety, plant health
- Cultural factors - plant density, plot slope, cover crop, mulch, residual nutrients of previous crop, availability of moisture.
- Climatic factors - temperature, sunlight, rainfall etc.

The deficient nutrients are supplemented to the soil through various organic and inorganic sources. Organic sources are usually bulky. They provide volume to the soil, improve the soil structure, improve the water holding capacity of the soil, provide optimum environment for beneficial soil microbes to be active.There are concentrated forms of organic manures also like bone meal, blood, oil cakes, urine etc. Organic manures are the right choice to counter the chemical toxicity in the soil and various gaseous emissions from agricultural fields due to application of chemical fertilizers.

Fertilizers are concentrated inorganic salts, nutrition specific supplement which is readily soluble and available in the soil, ready for plant uptake.

A. Important organic manures

1. Cattle manure or farmyard manure

The dung of farm animals, farm wastes like straw, bedding of animals etc. constitutes this category of manures. It takes a long time to decompose and be good enough to be used. The nutrient status of animal excreta depends upon the kind of feed and fodder given to the animal.Generally it contains 0.6% N, 0.35% phosphorus and 0.6% potassium. Dried and decomposed farmyard manure is applied as a basal dressing and incorporated into the soil during field preparation and ploughing.

2. Night soil or Poudrette

It is very rich in nitrogen. The problem is handling it because of smell and flies associated with it. Night soil when mixed with equal volume of ash and ten percent charcoal powder yields odourless produce containing 5.5% of N, 4.4% P_2O_5 and 2.0% K_2O.

3. Guano

It is bird droppings found on the coasts, mainly the coasts of Peru and South America. Since it is very rich in nitrogen (10.15%) and phosphorus (9.82%), it has to be mixed with soil in the ratio of 1:6.

4. Bone meal

It is made by grinding animal bones. This is rich in calcium, phosphoric acid and lime. Bone meal when steamed contains very high phosphoric acid (23%) and about 3.5% N.

5. Oil cakes

When oil is extracted from seeds like groundnut, sunflower, safflower, mustard,neem, sesame, castor coconut etc. the residue left is called oil cake. It contains 4% N and 1.5 -2% P. The Oil cakes are soaked in water and the liquid is used as crop supplement. The liquid acquired after soaking neem cake is used as pest repellent in crops.

6. Leaf mould

It is produced by slow decomposition of leaves and garden sweepings. The waste leaf material is heaped in a shady place or put in a pit, dug for this purpose and allowed to slowly decay by fungal action. It has high amount of humus and water holding capacity. The nutrient content depends upon the kind of leaf.

7. Wood ash

It is the residue, left after burning wood. It is rich in lime and potassium. Can be used in soil amendments in potted plants.

8. Coir compost

Coir pith is obtained as waste from coir industry. It is composted and brought to use as soil supplement and conditioner. It is composted by the fungus *Pleurotus sojarcaju.* Though it takes longer time to get composted, it is rich in plant nutrients. The percentage of N, P_2O_5 and K_2O is 0.26-1.06, 0.06, 1.2 respectively. It is has calcium and certain micronutrients also. It absorbs water to the extent of five times its weight, thus has very good water holding capacity.

9. Vermi compost

Vermi compost is made when earth worms are introduced in the organic waste material and animal dung. Some species of earthworms like the *Eudrilius eugeniae* are voracious feeders thus decompose the material very fast. The excreta of the worms is called worm casts, which is very rich in the nutrients required for the growth of plants.

10. Green manuring

Green manuring is the practice of incorporating green plant materials into the soil. Crops like Sunnhemp (*Crotolaria juncea*), Sesbania (*Sesbania speciosa*), Cowpea (*Vigna sinensis*) Gliricidia (*Gliricidia maculata,* Daincha (*Sesbania aculeata*), Pungam (*Pungamia pinnata*), Cluster bean *(Cyamopsis tetragonoloba*) etc. are grown in the field and are ploughed back into the field when very young, about one and a half to two months old. It should not be allowed to flower. Usually legumes are preferred as green manure crops. A 45 to 60 day old crops can enrich the soil to the tune of 60 to 90 kg N per ha. Apart from enriching the soil with nutrients, green manure crops also builds humus, act as soil conditioner and enhance the activity of beneficial microbes in the soil.

11. Biofertilizers

Biofertilizer, as the name suggests, is the nutrition supplied to the crop by introducing nutrition mobilizing microbes to the soil. The beneficial microbes are mixed with a carrier. The carrier keeps the microbe viable. These microbial carrier based preparation is viable for a limited period of time only, unlike the previously discussed organic manures and inorganic fertilizers. However, biofertilizers should not be mixed with chemical fertilizers since their contact with chemical fertilizers might result in killing the microbes as well. Bio fertilizers are very useful for organic cultivation and are used in crops in various ways.

- **Seed treatment**: For the seed required for one hectare, 400 g of bio fertilizers is required.
- **Seedling dip**: In case of bare root seedling transplantation, seedling dip is recommended. One kilogram biofertilizer is mixed in required quantity of water and roots of the seedlings required for one hectare are dipped in it for ten minutes before transplanting.
- **Basal inoculation**: Mixed in the soil during the field preparation, or in pits for fruit trees. Three kilograms of biofertilizer is broad cast uniformly in one hectare.
- **Spraying:** Used as sprays on standing crop at different stages.

Various beneficial soil microbes have been identified and categorized on the basis of the nutrient it mobilizes.

- Nitrogen fixing - *Azospirillium, Azotobacter, Rhizobium*
- Phosphorous mobilizing - *Pseudomonas, Microbacterium laevaniformans*.
- Potassium mobilizing - *Frateuria aurantia*, *Acidothiobacillus, Paenibacillus*.

Mycorrhizal inoculants

Mycorrhiza (AM, Arbuscular Mycorrhiza) is the outcome of the symbiotic relationship of phycomycetous fungi with angiosperm plant roots. It makes nutrients like N, P, K, Ca, S and Zn available to the plants.

B. Important inorganic Fertilizers

1. Nitrogenous Fertilizers

Nitogenous fertilizers are grouped into four kinds - Nitrate, ammonia, ammonium salts and amide.

S.No.	Nitrogenous fertilizers	Nutrient content (%)	Form of Nitrogen
1.	Sodium Nitrate	16.0	Nitrate
2.	Potassium Nitrate	12.5 – 13.5	Nitrate
3.	Ammonium Sulphate	20 - 21	Ammoniacal
4.	Ammonium Chloride	25 - 26	Ammoniacal
5.	Anhydrous Ammonia	99	Ammoniacal
6.	Ammonium Nitrate	33.0	17 - 18 % Ammoniacal, 17 - 18 % Nitrate
7.	Ammonium Sulphate Nitrate	25.6	19% Ammoniacal, 6.6% Nitrate
8.	Urea	46.0	Amide
9.	Calcium cyanamide	20.6	Amide

Urea and Ammonium Sulphate are the most preferred nitrogenous fertilizers. Urea is white in colour and crystalline, easily dissolved in water and cheap. Ammonium Sulphate is fine, crystalline and readily soluble in water. Prolonged use of Ammonium Sulphate increases soil acidity.

2. Phosphatic Fertilizers

These are grouped as natural, processed, chemical and by-products of phosphates. Popularly used phosphatic fertilizers are:

i) Superphosphate

Available in three grades

a) Single Super Phosphate - it has 16 - 20 % phosphoric acid.

b) Dicalcium Phosphate - it has 35 - 38 % phosphoric acid.

c) Triple Super Phosphate - it has 44 - 49 % phosphoric acid.

ii) Basic slag

It is a by-product from steel factories. It has 6-20% phosphoric acid; it is alkaline in nature, hence good for acidic soils.

iii) Rock phosphate

It is the natural deposit of rock in the USA, Poland , Morocco, Brazil and some islands in Indian and the Pacific ocean. It contains 25-35% of phosphoric acid, which is insoluble in nature. Good for acidic soils.

3. Potassic Fertilizers

Commonly used potassic fertilizers are potassium chloride or muriate of potash and potassium sulphate or sulphate of potash. Potassium chloride is crystalline and contains 50 to 63% K_2O. Potassium Sulphate has 48-52 % K_2O. It improves the quality of the produce.

4. Mixed fertilizers

Mixed fertilizers supply more than one nutrient. When a mixed fertilizer contains nitrogen, phosphorous and potash, it is called as complete fertilizer.

Advantages

- Saves time in application and consequently labour man-days are also saved.

Disadvantages

- All nutrient needs of a crop cannot be satisfied with one mixed fertilizer
- Costlier

5. Liquid fertilizers

Liquid form of fertilizer is available for a single nutrient and combination of nutrients. Liquid fertilizers are expensive than the other forms. They are used as foliar sprays or for fertigation in drip system of irrigation.

Choice of nutrient source

Inappropriate choice of nutrient source, specially chemical fertilizers, greatly affects the nutrient utilization and uptake. Some nutrients may get fixed and immobilized, when some other is high in concentration. Reaction of fertilizer with soil pH and residual nutrient status should be well analysed.

Time of application of fertilizers and manures

The time of fertilizer application is crucial. The nutrient should be available to the crop when the crop needs it. For example, plants that flower in February / March need fertilizer application in October/ November. A substantial dose of fertilizer should be applied at the time of field preparation.

Nitrogen is an important mineral nutrient which is required throughout the phase of plant growth and development, so a regulated input, in split doses, spread across the phases of plant growth is required. Too much nitrogen application at one time is also not advisable since nitrogen is easily dissolved in water; it is easy to leach beyond the absorption zone of plant feeder roots. Phosphorus should be applied before planting or sowing because it is required mostly for the initial growth and development of the plant, more so because it is released slowly.

Potassium, like nitrogen, is required throughout the plant growth, but it is a slow release nutrient, hence it should majorly be applied as basal dose.

Method of application of fertilizers and manures

1. Broadcast method

The fertilizer is broadcast on the entire field during field preparation, just before the final ploughing. A basal dose consisting of calculated amount of nitrogen, phosphorus and potash, required for a crop may be incorporated. Total required dose of phosphorus and potash and a part of nitrogen is applied as basal dose. The total amount of Farm Yard Manure may also be incorporated as basal.

Broadcast application is done while the crop is standing in the field; this is called 'top dressing'. Since nitrogen is required in split doses, mostly top dressing is done with nitrogenous fertilizers.

2. Band application

This method is followed in annual crops which are grown in ridges and furrows e.g., carrot. A band of fertilizer is placed on one side of the ridge. For perennial fruit trees, top dressing is done by making a six inches wide shallow circular band, just below the canopy spread, because this is where the feeder roots are located and nutrient uptake is efficient.

3. Seedling dip or Starter solution

Seedlings are dipped in bio fertilizer or mild liquid fertilizers (starter solution) before transplanting. This improves the seedling survival and plant stand.

4. Seed treatment

Seed treatment with biofertilizers and soaking planting material like bulbs and tubers in mild nutrient solution are found to be effective.

5. Foliar application

Foliar application is done at various stages of plant growth, mainly in emergency situation. The liquid fertilizers or even a mild spray of urea is applied to get instant effect in plant health. When nutrients are applied as foliar sprays, uptake is very efficient. To correct the deficiencies, micronutrient sprays are given.

When foliar sprays are given with liquid fertilizers, care should be taken to use them judiciously, because they are expensive. Erratic use of liquid fertilizers will increase the cost of production.

When spraying urea on foliage, care should be taken to make the solution of the right dosage. If the urea solution is high in concentration, the foliage will burn and cause urea injury.

The foliar spray should be given late in the evening. Early morning spray will mix with the dew and reduce the concentration, while afternoon spray will dry quickly, thereby scortching the leaf.

6. Fertigation

Liquid fertilizers are used in drip irrigation systems. This type of fertilizer application is called fertigation. Small doses of liquid fertilizers are fed to the crops along with irrigation water at regular intervals.

Integrated Nutrient Management

Integrated nutrient management (INM) or Integrated plant nutrient supply (IPNS) system aims at maintaining soil fertility so that the required amount of nutrients are available to the plant, throughout its growth period and optimum yield and productivity are achieved in a sustainable manner. This system takes into account all possible sources of nutrients supply and strikes a strategy to obtain maximum benefit in an integrated manner. This system abstains from using too much chemicals for pest and disease control as well as in fertilizer form as the key inputs are bio fertilizers, bio inoculants, organic manures and sustainable practices for pest and disease control. This system conserves the biological and physical properties of soil. This is extensively used when horticultural crops are cultivated using GAP (Good Agricultural Practices).

8

Irrigation Management

Water is an extremely essential component in plant growth and development. It is of utmost importance for photosynthesis, cell division, cell growth / elongation, respiration, nutrient uptake, mineral utilization and much other metabolic process, which directly affect the quantity and quality of the produce. Hence quantity and quality of water play a vital role in performance of a crop. A well distributed water supply, throughout the growth of plant is essential. Rainfall is the major source of water for the crops. Rainfall recharges the surface water resources such as rivers, streams and ponds and also the ground water resources like wells.

There are many advantages of irrigation

- If a land has perennial source of enough irrigation water, the grower has flexibility to grow a wide range of crops.
- The plant response to fertilizer and other production and quality enhancing inputs is better. The yield and quality of the produce is superior thus better marketability and better rates can be expected.

Irrigating field more than required quantity causes water logging, fungal infections like damping off and the crop becomes susceptible to pests.

The time when a crop is to be irrigated and the quantity of water the crop should receive depends on a number of factors as enumerated below:

1. Annual rainfall

This greatly influences the irrigation schedule. Rainfall should be adequate and distributed around the year to cater to plant requirements. Unfortunately, rainfall is now erratic and undistributed, some parts of the country face drought while other parts may see flood, and both these extreme conditions are detrimental to crops. A good rainfall in a year stores good amount of water for all the year round. In case of low rainfall, that is when available water is less, the irrigation schedule gets limited to only critical stages of the crop. Sometimes there is drought and the water is only sparingly available and this negatively affects the production.

2. Nature and structure of crop

The shallow rooted crops need frequent irrigation while the deep rooted plants manage the required moisture if ground water is sufficiently charged. Drought tolerant plants need to be planted in areas where soil moisture availability is less. In areas where water is copiously available and papaya plants like bananas may be planted.

The amount of water the roots absorb is proportional to the transpiration rate of the plants. Plants with leaves having larger lamina transpire more, thus water requirement of such plants is more. The plants with faster growth, like greens and bamboo too have a higher water requirement.

3. Plant growth stage

The plant requires different amount of water in different stages of growth. Plant establishment, vegetative growth and fruit development phases are crucial. In case of deciduous perennial tree crops, when the plants slip to dormant phase, the water requirement is almost zero.

4. Moisture stress

The dry months of the year cause moisture stress, sometimes to the point of irreversible permanent wilting. Irrigation is required when leaf start to droop. The exact moisture status of the soil can be determined by tensiometer, neutron probes, and many more methods. These help to give a regulated supply of water to the plants thus increasing the water use efficiency.

5. Soil quality and texture

Deep soils, fine textured soils and the soils with high organic matter hold water for a long time. Clay soil holds water for a longer time than sand.

Types of irrigation systems

A. Surface irrigation

The irrigation water is let on the soil surface and allowed to spread through gravity. In all the methods of surface irrigation, there is wastage of water since the water is taken from source to the field by open irrigation channel and in this process, there is wastage of water. Moreover, the area near the channel may get excess water while the area away from the channels may not get sufficient water if the field is against gravity or if water speed is not adequate. For surface irrigation it is important for water source to be at a higher point than the field and further fields should be along the gradient of gravity. Surface irrigation is carried out in different ways -

- **Flooding:** This method of surface irrigation is a very old method, it is still widely followed in wetland cultivation.There is wastage of water in this system and more water is required in this system of irrigation, as the water spreads in all the area, not just where the plants are, consequently a lot of weeds sprout.
- **Check:** It is like flooding, but check bunds are made to enclose a block of crop. Here channels are provided to help water reach the total area.
- **Beds:** The water from the main irrigation channel is allowed in the beds of direct sown crops like the greens. There is a possibility of the shallow sown seed to be dislodged with water current and get collected in the deeper parts of the bed, hence it is imperative to have levelled beds.
- **Furrows:** The water from the main channel is allowed to pass through furrows, made for the crops which are grown in ridges like onions, radish, carrots etc.
- **Basins:** Basins are made around the tree trunk and joined to the main water channel through a small channel. This method of irrigation is widely practiced for perennial trees like mango, litchi, coconut, sapota, moringa etc. The water from main irrigation channel gets diverted to the basin, after filling the basin, it comes back to the main channel.
- **Ring:** The bunds are made around the tree trunk as rings and they are all connected by irrigation channel. This system is mainly used to irrigate cucurbits like pumpkin, ashgourd etc.

B. Sub soil irrigation

The irrigation system is laid under the soil surface. The weed growth is less.

Pipe method

This system requires laying of pipes underground, so that water can be supplied to the root zone of the crops. Holes are made in the pipes according to the distance between the plants. Since this system is elaborate and high investment in material and labour is required. This kind of system does not allow deep ploughing. However, efficient water usage is there, since water is not wasted by evaporation and overflowing.

Pitcher method

A water filled pitcher is used to irrigate the root zone. This system is particularly efficient to irrigate young trees in drought prone areas. A pit of approximately 50 cm depth and 50 cm diameter is dug under a tree, without damaging the roots. A 20 litres pot with a few small holes at the base is firmly placed in it. The water filled pitcher slowly releases water to the root zone of the tree. The pitcher needs to be refilled every 4-5 days. Water is efficiently used.

C. Other irrigation systems

1. Sprinkler irrigation System

A kind of surface irrigation system, also known as overhead irrigation system. Pipes are laid on the surface and sprinklers are fitted at intervals. The sprinklers rotate and irrigate the area. Good for large lawns, vegetables and for cultivation in terraces, where surface irrigation cannot be done due to possibility of soil erosion. The system installation cost is high.

Advantages

- Does not disturb soil hence minimizes soil erosion.
- No manpower is required for irrigation.
- Suited for terrace cultivation in crops like tea, vegetables and flowers cultivated in the hilly slopes.

Disadvantages

- During high day temperature, scorching of leaf lamina may occur when water droplet touches the hot leaf surface.
- On a windy day, there is water drift and the irrigation will not be uniform.
- May write pathogens, when water droplets stays on the leeaf surface.

2. Drip irrigation

Like the sub soil irrigation systems, this system also targets the root zone. Pipes are laid on the soil surface and water drips at the collar zone of the plant, through holes in the pipe. The water then trickles down to the root zone, hence this is also called 'trickle irrigation'. Daily supply of water as per the plant requirement is ensured. The system is very technical and requires high investment. In high value green house crops, drip irrigation system is computerized.

A typical drip irrigation system has the following components

- A water supply source with pump – used to distribute water to the field.
- Filters near the water source - to filter the water and remove suspended impurities, since these impurities will clog the pipes.
- Main line - for conveying water from pump set to the field.
- Lateral pipes - It is fitted to the main water line. It helps deliver water to the rows. The position of fixing laterals is flexible. It can be fixed according to row to row distance available in the field.

- Fertilizer mixing tank – water soluble fertilizers are mixed with irrigation water and supplied to plant and the method is called fertigation, which is discussed in detail at the end of this chapter.
- Drips or emitters – fitted in the laterals at plant to plant distance, so that water can be delivered to the plants.
- Control system – it mainly consists of water meter (to check the quantity of water) and pressure gauge (to measure the pressure at which water is flowing).

Advantages

- Water is used very efficiently, without wastage.
- Water is supplied as per the needs of the plant at the time required by the plant and in optimum quantity, thus plants are healthy and yield better.
- Less manpower is required since the drip lines are already laid, one person is enough to switch the system on and off. No extra manpower is required to apply fertilizer as fertilizers are applied along with irrigation water.
- Fertilizers are used efficiently and delivered at root zone to feeder roots, thus plants are vigorous and plant stand is better.
- Since the nutrients and water are only applied to the root zone, weed growth is minimal.

Disadvantages

- High initial investment.
- Cost of water soluble fertilizers is high.
- Clogging of drippers due to salt deposition.

Fertigation

Application of plant nutrients along with irrigation water, through drip lines is called 'Fertigation'. The success of fertigation depends upon:

- Calculating the exact quantity of fertilizers for each growth stage.
- Precise delivery of exact quantity of fertilizer, may be daily, once every alternate day, or once weekly.
- Use of water soluble fertilizers which are compatible to each other. The fertilizers should work best in the available soil type and water quality. Common water soluble fertilizers are Potassium Nitrate, Mono Ammonium Sulphate, etc.

Mixing of incompatible fertilizers like Ammonium Sulphate with Potassium Chloride, Calcium Nitrate with Phosphates or Sulphates, Magnesium Sulphate with Mono or di-Ammonium Phosphate, Phosphoric Acid with iron, zinc, copper or Manganese Sulphate, is not advocated.

Common fertilizers like Urea and Muriate of Potash can also be used but they should be dissolved in the tanks one day prior to application.

In Banana, studies revealed that fertigation with 75-100% of recommended dose of fertilizers, at weekly intervals resulted in increased bunch weight and fruit quality.

Fertigation schedule at weekly interval for many fruits, vegetables species and flower crops have been standardized

Advantages of Fertigation

- Saving in cost of manpower in fertilizer application.
- Nutrients are used efficiently as they are supplied in exact doses as per plant requirement.
- Nutrient loss through leaching and fixation is reduced.
- Increased yield and quality of produce.

Water is a precious natural resource, that has been misused. Water should be used cautiously in growing crops. This includes proper management of water by reducing wastage in channels, runoff, spilling, seepage and evaporation. Mulching should be done to conserve soil moisture. Crops which demand less water, with no compromise in yield and quality, should be chosen where ever possible.

In recent days, it is imperative to adopt fertigation systems in the new concepts of High density planting, protected cultivation and precision forming techniques.

9

Weed Management

Different crops are grown in a field. Any plant growing in a place where it is not desired competing with the main crops, has to be weeded out and such plants are called 'weed'.

Effect of weed menace

The weeds

- Reduce crop yield and production efficiency. It competes with the main economical crop for space, water and nutrients.
- Reduce crop quality.
- Hamper intercultural operations.
- Reduce crop quality.
- Obstruct the flow of water in the irrigation channels.
- Exert allelopathic effect on crops.
- Harbour insect and pests, which affect the main crop of interest.
- Increase the cost of production.

Kinds of weeds

Weeds can be classified

- On the basis of cotyledons in the seeds as monocot or dicot.
- On the basis of the time it takes to complete its lifecycle as annual/biennial/ perennial.
- On the basis of leaf, as broad leaf or narrow leaf.
- On the basis of the root depth as shallow rooted or deep rooted.

Methods of weed control

Weeds can be controlled by

- Elimination at entry level by getting clean propagules and pre-curing the inputs like the farm yard manure. Not allowing leaf eating animals inside because their excreta contain seed, which germinate and become menace.
- Eradication by way of chemicals.

Principles of weed management

Preventive weed management

- Use weed free seed.
- Use organic manure only after complete decomposition to kill weed seeds. Undecomposed manures are the major sources of weed seeds in the field.
- Clean the tillage implements after tilling in weed infested areas.
- Avoid incorporation of soil from weed infested area.
- Inspect nursery for vegetative propagules.
- Before planting a crop, remove weeds from non-crop land such as irrigation canals, fence,etc.
- If a weed is seen immediately remove it before it flowers, this prevents further proliferation of weeds.
- Use weed seeds to filter irrigation water.
- Restrict livestock movement in the field.

Weed management through agronomic practices

- State seed bed – the seed beds are formed about a week before planting the main crop. Moisture is provided and all the weed seeds in the area germinate. These then can be removed manually or chemically.
- Land preparation – when the perennial grass weeds are dominant, it is best to plough deeply in such a manner that the roots, rhizome and other propagules 0of the weed, which are under the soil are exposed and dried by wind and sun. In case of annual weeds, shallow cultivation is desirable.
- Crop rotation – rotation with fast growing crops with dense canopy is effective in smothering the weeds.
- Manual weeding – it is a tedious process but very effective if the weed is uprooted along with the root.
- Irrigation – weed intensity is low in drip irrigation system. Pre sowing irrigation is very effective since the weed seeds germinate, which die when not irrigated again.
- Maintaining optimum plant stand – proper plant stand should be maintained so that there is no empty space in the field for growth of weeds.
- Mulching – mulch material like plastic sheets and dry plant material is spread before the emergence of the weeds. Mulching may be supplemented with hand weeding to achieve effective control.

- Solarization – it is a method in which plastic sheets are spread on moist soil to trap soil heat. Soil solarisation increases the soil temperature by 8-10°, which effectively controls the weeds.

Weed management with chemicals

The chemicals used to kill weeds are called as herbicides and weedicides.

- Soil fumigants – Soil fumigation is done with methyl bromide, ethyl alcohol, 2 – 3 weeks before sowing seeds. This is very effective in controlling weeds.
- Foliar spray – Specific chemicals are used to kill monocot and dicot weeds. Some chemicals are available, which kill the weeds indiscriminately.
- Herbigation – Herbicides or weedicides are applied with surface irrigation water as well as through sprinklers.

Herbicides are classified as follows

I. **On the basis of seed germination and emergence –**

a) Pre-emergent- This kills weed seeds before germination e.g. Diuron

b) Post emergent - This kills the weed plant e.g. 2, 4 – D, glyphosate. Post emergent herbicides are further classified as –

i) Contact herbicide – kills the plant when it comes in contact with the plant e.g. Paraquat.

ii) Translocated – Absorbed by plant system, takes time to kill the weed e.g. Glyphosate

II. **On the basis of selectivity –**

a) Selective herbicide – acts on monocots or dicots e.g. Dalapon for monocots and 2,4 – D for dicots.

b) Non selective herbicides – kills both mono and dicot plants e.g. Glyphosate.

Benefits of herbicide use

- Early and efficient weed control.
- Suited for crops at close spacing, where other methods of weed control are not possible to follow.
- Controls weeds without any crop injury.
- Very useful when manpower is scarce.

Precautions when using herbicides

- Spray when optimum moisture is available.

- Use the correct herbicide in proper dose and time.
- Use clean water to mix herbicide.
- Spray during morning or evening hours.
- Do not mix other plant protection chemicals with herbicides.
- Avoid spraying in areas with high wind velocities, to avoid chemical drift.
- The person who is spraying should wear protective gear like gloves, goggles and masks.
- Use hoods to avoid drifting of chemical, not to touch the main crop.

To achieve best results with herbicide

- Choose the type of herbicide, according to the available weed flora.
- Time of application should be such that the herbicide efficiency is maximum.
- Method of application
- Optimum dosage – lower dose creates resistance in weed and high dose leads to chemical wastage.

To help make a discrete choice, below is the list of important herbicides, their properties and dosage -

Herbicide	Properties	Dosage
2,4 – D Di chlorophenoxy acetic acid	Selective and translocated Intercepts cell growth, leaves droop, respiration increases and the weeds die due to exhaustion of starch reserve. Requires 8 hours of clear weatherfor best performance.	31g/10 lit with 5 ml wetting agent or surfactant.
Oxyflorofen	Pre – emergent, non-selective	20 ml in 10 lit water with 5 ml wetting agent or surfactant.
Paraquat di chloride	Non – selective, contact, needs one hour of clear weather to act. It acts by decomposing chlorophyll	25/33/50 ml in 10 lit with 5 ml wetting agent or surfactant.
Glyphosate N – phosphomethane glycine	Non – selective, translocated, interferes with synthesis of chlorophyll, amino acids and protein.	25 – 30 ml / 10 lit water with 5 ml wetting agent or surfactant.

Wetting agent or surfactant

It is used to help the water based herbicide spread properly on the surface of the weed and adhere to it. Addition of surfactant to water brings its surface tension from 70-75 dynes/cm^2 to 40-60 dynes/cm^2. Triton A is a commonly used surfactant.

Weed Management with Biocontrol Agents

This involves using natural enemies of weeds to control weeds.*Crocidosema lantana* moth and *Teleonemias crupulosa* bug feed on *Lantana camera* weed and kill it naturally. Cactus *Opuntiadilleni* is controlled by scale insect

(Dactylopiusto mentosus). Water hyacinth *(Eicchornea*) is controlled by hyacinth weevil (*Necochetinaeichhorniae*). Natural compounds or allelopathic substances released by some plants inhibits the growth of some weeds. When fungi, bacteria or virus are used to control weeds, they are called bio herbicides.

Fungus *Phytophthora sp* is used to control *Morreniaodorata* vine in citrus gardens in USA.

Integrated Weed Management

No single weed control method is adequate to control weeds. A combination of different methods in different cropping systems results in effective weed control. This approach is called integrated weed management.

For effective weed control, use of herbicides may be integrated with cultural methods of weed control. Similarly soil solarisation alone is not effective to control *Cyperus rotundus*, thus it may be integrated with hand weeding. *Orobanche sp* of parasitic weed cannot be controlled by herbicides alone. In orchards, where tree to tree spacing is wide, intercropping makes weed control efficient in the initial years of establishment. This practice does not leave any free land for weed growth. Integrated weed management practices may be different for different crops. Weed management practices that can be integrated in a particular area and crop depends on the type of soil, kind of area, availability of labour, manoeuvrability of machines, type of crop – annual or perennial, type of crop culture – mono culture or multiple cropping and the choice of the grower.

10

Propagation Techniques

Plant propagation techniques refer to the methods by which the plants are multiplied and increased in number.

Methods of Propagation

i) **Sexual Method of Propagation** - Seeds are used as propagating material.

ii) **Asexual Methods of Propagation** - Plant parts, other than seeds are used for plant propagation.

Sexual method of propagation

Advantages

1. For evolution of new varieties through breeding, the hybrids are raised from seed.
2. This is the only option, other than tissue culture, to propagate plants that cannot be propagated by asexual methods, e.g., papaya.
3. Most of the rootstocks are raised from seeds.
4. Seeds can easily be transported from one place to another.
5. Seed propagated rootstocks are hardy and develop better root system.
6. Occurrence of polyembryony (more than one embryo in seed) in citrus and mango leads to the development of uniform seedlings, giving advantage of asexual method.
7. Self pollinated vegetables can be easily and rapidly propagated through seeds, as it is easy to maintain purity.
8. Multiplication is easy in certain ornamental crops, which produes seeds abundently.

Disadvantages

- Seedlings have a long juvenile period and come into bearing later as compared to asexually raised plants.
- Due to heterozygosity segregation of characters, the progeny is not true-to-type.

- Seeds loose viability over time.
- Nurseries are required to raise the seeds for transplanted crops.
- Directly sown seeds have irregular plant stand.

The seedlings for transplanted crops should be raised in nurseries, before they are transplanted in the main field.

Types of nursery

1. **Temporary nursery:** It can be changed from one place to another, depending on needs. It has no permanent structure involved. This includes the flat bed nursery, raised bud nursery and nursery raised in protrays.
2. **Permanent nursery**: An area is enclosed with sidewalls and drainage holes.The walls are constructed with concrete to a height of 75 cm. Seeds are sown in soil enclosed in the concrete structure. After removal of each batch of seedlings, the soil is enriched with manures, because there is a constant removal of nutrients, when seedlings are grown. This is a permanent structure exclusively built to raise a nursery. A separate area is allocated for this purpose, which cannot be used for any other purpose.

Advantages of raising seedlings in nursery

- It is very convenient to look after the tender seedlings.
- It is easy to protect the seedlings from pests and diseases.
- Economy of land usage (duration in the main field is reduced)
- Valuable and very small seeds can be raised effectively without any wastage.
- Uniform crop stand in the main field can be maintained by selecting true to type healthy, uniform and vigorous seedlings in the nursery itself.

Preparation of nursery beds

Selection of site

- The nursery area should be near to the water source.
- Generally, the location should be partially shaded, may be under the trees. If not, artificial shade has to be provided.
- Should be well protected from animals.
- Proper drainage facilities should be provided.
- Should be nearer to the office for effective supervision.

Selection of soil

- A medium textured loam (or) sandy loam soil is preferred.
- Soil should be rich in organic matter.
- Soil depth should be preferably 15-25 cm.

Types of nursery bed

1. Flat bed
2. Raised nursery bed
3. Raising in protrays

Flatbed nursery

- A site is chosen and tilled to a fine tilth. Stones, weeds and all inert matters are removed.
- Two parts of fine red earth, one part of sand and one part of farm yard manure (FYM) can be incorporated to each bed to improve aeration and fertility of the soil. The soil is drenched with 4% formaldehyde or 0.3% copper oxychloride, to kill the pathogenic spores in the soil. Soil solarisation with white polythene sheet may also be done to check the nematode infection and weed growth.
- Beds are marked and leveled.
- Beds are demarcated by making ridges and seeds are sown in the beds.
- It is the simplest way of nursery bed making.
- In this kind of bed, if the level is not even, the water stagnates at the lowest point during irrigation and there is a possibility of the seeds being dislocated and getting collected at this point.
- Overcrowding of seeds at one point may lead to mortality.

Raised nursery bed

- A site is chosen and tilled to a fine tilth. Stones, weeds and all inert matters are removed.
- Two parts of fine red earth, one part of sand and one part of FYM can be incorporated to each bed to improve aeration and fertility of the soil. The soil is drenched with 4% formaldehyde or 0.3% copper oxychloride, to kill the pathogenic spores in the soil. Soil solarisation with white polythene sheet may also be done to check the nematode infection and weed growth.

  Raised beds are demarcated and made. Height of the raised bed should be 10-15 cm with a width of 1 m and length may be according to the requirement and convenience. The width is kept at one meter so that the weeds can be removed from both the sides of the bed conveniently and also the plant inspection is proper.

Advantages of raised nursery bed

- Water movement will be uniform and drainage of excess water is possible.
- Germination percentage of seeds is normally high. Since there is no water stagnation.
- Operations like weeding, seedling health monitoring and plant protection measures are easy.

Raising in protrays

This method is used for raising high value hybrids for commercial purpose. The seedlings are produced under protective structures such as insect proof shade net house and low cost naturally ventilated greenhouses.

Protrays

- Protrays (Propagation tray) are also called as seedling trays or plug trays.
- The most commonly used are 98 celled trays for raising hybrid seeds of vegetables and flower.
- The trays are generally 54 cm long and 27 cm wide and the depth of each cavity is 4 cm. These trays are made of polypropylene and are re-usable.
- Life of the tray depends on the handling; never the less it lasts at least 2-3 seasons.
- Seedling trays have been designed in such a way that a sapling gets a pre-calculated growing media and the right amount of moisture as the trays have pre-punched holes to each cavity for proper drainage of excess water and also right spacing.
- Use of these trays makes the long distance transportation of seedling convenient.

Media for protrays

1. The most common media used in protrays is digested coir pith compost (also called as coco peat, bio compost). It is steam sterilized to prevent nursery diseases. It is enriched with required nutrients before use.
2. Digested coir pith compost with equal proportion of either vermicompost or sand has also given good result in protrays.

Method of raising seedling in protrays

- The digested coir pith is lightly moistened and filled in the tray plugs.
- One seed per cell is sown and covered with media.
- Since the media is already moist, no irrigation is required after seed sowing.
- Ten trays are kept on one over the other and covered with a polythene sheet and kept undisturbed for three days. This conserves moisture and provides adequate warmth for the seed germination.
- Seeds start emerging after about 3-6 days of sowing depending upon the crops. Then the trays are kept spread over a bed with polythene sheet.
- The germination trays are then lightly irrigated with fine sprayer or rose can, if moisture is not noticed when media is touched.
- The trays are also drenched with fungicides as a precautionary measure against seedling mortality.
- The media may be supplemented with nutrient solution if the seedlings show deficiency symptom. Spray 0.3% of 100% water-soluble fertilizer (All 19:19:19 with trace elements) twice (12 & 20 days after sowing).
- Care must be taken for spreading the trays when the seedling is just emerging, otherwise seedlings will get etiolated.
- Protection of the trays from rainfall is important. This may be ensured by covering the protrays with a polythene sheets in the form of low tunnel.
- The seedlings will be ready in about 21-42 days for transplanting into the main field depending upon the crop.
- The seedlings should be hardened and acclimatized before transplanting in the main field, so that the survival percentage is high. This is done by withholding irrigation and reducing the shade before transplanting.
- Vector spread viral diseases, hence are a threat. Systemic insecticides like Imidacloprid (0.02 ml/l) 7-10 days after germination and before transplanting should be sprayed for managing vectors.

Seed dormancy

It is a condition where the seed does not germinate when sown in normal conditions.

Physiological reasons and methods of breaking dormancy

1. Hard seed covering impermeable to water, also called seed coat dormancy. In this condition, the seeds are not able to absorb water and germinate due

to the presence of hard seed coat. Physical removal of seed covering is followed for breaking the dormancy.

2. Hard seed covering resistant to embryo expansion. In this condition, the hard seed covering does not allow the embryo to expand; expansion of embryo is a normal process after the seed imbibes water. Here, either the embryo dies or experiences delayed germination. Example: walnut and certain stone fruits. To help the seed germinate, the seed shell is removed before soaking in water.
3. Seed covering containing germination inhibitors. In this condition, the seed coat, seed shell or any other seed part may be involved in producing the inhibitors for germination. This can be overcome by soaking the seeds in water and leaching the inhibitors.
4. Seeds with rudimentary embryo- In this condition, the embryo size are very small, under developed and do not germinate. These seeds are generally the floaters which are light and non-viable.
5. Seeds with internal dormancy-The inner tissues, tissues enclosing endosperm and the inner integument layer of the seeds have inhibitory action in seed germination. Treating the seeds with chemicals like KNO_3 and growth regulators like GA_3 may help to overcome the dormancy.
6. Physiologically shallow dormancy- This is temporary dormancy which is present mostly in fleshly harvested seeds. These seeds germinate after they are stored for certain period of time after extraction.
7. Physiologically intermediate dormancy- This dormancy is seen in conifers. Moist chilling is required to encourage germination. To break the dormancy, the seeds may be kept in moist sand before germination. Vernalization can also be done.
8. Double dormancy- In this condition, seed coat dormancy and internal dormancy both occur together. Seeds in this condition require more time to germinate.

Seed treatment: It is a process of application of chemicals or protectants (with fungicidal, insecticidal, bactericidal or nematicidal properties) to seeds that prevent the carriage of insect or pathogens in or on the seeds.

Objectives of seed treatment

1. **To control disease:** There are some seed borne or soil borne diseases. Seeds are treated with fungicides like Captan or Thiram (2g/ kg seed), or Carbendazim (1g/ kg seed) or *Trichoderma viride* (4 g/ kg of seed), 24 hours before sowing to control the seed borne pathogens. Bio fertilizers like

Azospirillum and Phosphobacteria can be mixed with rice gruel @ 250 ml per packet of *Azospirillum* or Phosphobacteria and dried under shade before sowing. Normally two packets (400 g) are needed for treating the seeds required for one hectare. These inoculants are helpful in getting healthy vigorous seedlings in the nursery itself so that the correct population can be maintained in the main field.

2. **To have convenience in sowing:** Difficulties are encountered in sowing certain crops due to special characteristics of the seed. e.g., coriander seed is to be split by rubbing it against hard surface. Seeds of amaranthus are mixed with fine sand or soil.
3. **To have quick germination:** Germination of certain leguminous crops is delayed due to thick seed coat which restricts water absorption. Such seed coats are broken to some extent by mixing them with coarse gritty sand and trampling or pounding it lightly in a mortar with a wooden pestle.
4. **To increase nitrogen fixation in legumes:** Legume seeds are inoculated with a particular *Rhizobium* culture. This is mixed with jaggery solution and applied to seed and dried in shade. It increases nodulation and thereby nitrogen fixation.
5. **To protect the seed against insect pests:** There are some insect pests like ants, white ants, in the soil which attack on seed and eat. To avoid this, seeds are treated with repellents.
6. **To induce earliness (vernalization treatment):** The seeds are exposed to cold temperatures. This stimulates production of gibberellins, which break the seed dormancy
7. **To induce variation:** Seed is treated to induce variation in its morphological and general structure by X- ray treatment.
8. **To break dormancy:** Some crops are having seed dormancy in fresh harvested produce. E.g. Thiourea 1% treatment to potato tubers.

Seed rate: (g/hectare)

Cabbage	Variety	-	650g	Tomato	Variety	-	400-500g
	F_1		300-550g		F_1	-	120-150g Indeterminate
Bellary onion		-	10kg			-	175g Determinate
Cauliflower	Variety	-	375g	Chillies	Variety	-	1 kg
	F_1		300-350g		F_1		200-250g
Capsicum	Variety		1250g	Brinjal	Variety	-	370-500 g
	F_1		375g		F_1		100-200g

Sowing of seeds in nursery beds

- Straight lines are drawn at a spacing of 10 cm to a depth of 1-2 cm, along the width of the nursery bed.
- Seeds are sown in the lines and covered with sand or fine soil or powdered FYM.
- Line sowing of seeds facilitates easy weeding.
- Depth of sowing determines the rate of emergence. If it is too shallow the seeds come up and dry out early. If it is too deep, the seedling emergence is much delayed, sometimes die without emerging. So a thumb rule is followed. Sow the seeds approximately at a depth of 3-4 times the diameter of the seed.
- After sowing the seeds, the bed is covered with paddy straw and watered by using a rose can so as to avoid packing of the soil and washing away of seeds.
- Watering should be done twice daily till the seeds germinate followed by once daily.
- The straw cover should be removed after germination.
- A week before transplanting, the seedlings should be exposed to full sunlight and the number of watering should be reduced so that the seedlings become hardy to bear the shock of transplanting in the main field and the mortality rate is less in the main field and proper crop stand is ensured.

II. Asexual Propagation

In this method of propagation, new plants are obtained from all the vegetative portions of the mother plant other than seeds.

Advantages

- In some fruit plants like banana, which do not bear seeds, this is the only method of propagation.
- The plants are just like the mother plant, true to type, uniform in growth, yielding capacity and fruit quality.
- The progeny has short juvenile phase, thus come into bearing earlier than plants from seeds.
- The advantages of rootstocks can be obtained by budding or grafting susceptible varieties on resistant/ tolerant rootstocks.
- Plants are uniform in growth thus uniform cultural practices can be followed and harvesting in also uniform and easy.

Disadvantages

- New variety cannot be evolved by this method.
- Plants are not as vigorous and long-lived as the seedling trees.
- Germplasm conservation requires lot of space and is expensive as compared to storage of seeds.
- Expensive when procedures like budding and grafting has to be done.
- Skilled labourers are required to perform grafting/budding.

Asexual methods involve propagation by

A. Cutting

B. Layering

C. Budding

D. Grafting

E. By specialized plant parts

F. Micropropagation

A. Cuttings

This is the easiest method of propagation. The process of propagation of plants by cuttings is known as cuttage. A cutting is a detached part of a plant that will produce roots when put in soil media and eventually produces a new plant which will be just like the parent plant. A cutting may be a piece of stem, a leaf or part of a leaf or a piece of root.

Types of cuttings

Cuttings are usually of three types, according to the plant part used.

i) Stem cuttings

ii) Leaf cuttings

iii) Root cuttings

i. **Stem cuttings:** Stem cuttings can be divided in to 4 types based on the degree of maturity and lignification of wood used in making cuttings.

   - Hard wood stem cuttings
   - Semi hard wood stems cuttings
   - Soft wood stem cuttings
   - Herbaceous cuttings

Hardwood cutting

Semi hardwood cutting

Softwood cutting

Hard wood stem cuttings

These cuttings are made from mature stem which has lignified. Hard wood stem cuttings of Grape, Fig, Pomegranate, Bougainvillea, Acalypha, Rose etc are made successfully.

Hard wood stem cuttings are of three types: **Straight or simple cutting**, **heel cutting** and **mallet cutting.**

Straight or simple cutting: It consists of only the current year's wood and doesn't bear any older wood. **E.g**. Hibiscus, Nerium.

Heel cutting: *A* small piece of older wood is retained at the base of each cutting E.g. Rose

Mallet cutting: An entire section of the older wood is retained. E.g. Thuja.

Straight

Heel

Mallet

Preparation and planting

- A disease free and healthy, about pencil thick shoot is selected from a desired plant.
- All the leaves are removed without damaging the axillary buds.
- A 20 – 30 cm cutting is made, which has 3-4 nodes. Each node will have an axillary bud.

- For identification, the lower end of the cutting is given a slant cut.
- Care should be taken when the bottom slant and upper straight cut are given. The cuts should be clean with a sharp knife. The upper cut is about 2.5-3 cm above the node, while the lower slant cut is about 1 cm away from the node. This is a very important point in the success of rooting of cuttings. The upper cut will be exposed to light and air and there are chances that it will dry a little initially, which is why the upper cut is away from node, so that when the tip is drying a little, the upper bud is not lost. The lower cut is embedded in soil and roots are expected from there. For rooting, the bud releases the rooting hormone to the cut part, that is why the lower cut is closer to the bud so that rooting hormone from bud reaches the cut portion faster and forms roots.
- In case of some difficulty in root cutting, the bottom cut is treated with rooting hormone so that rooting is induced.
- The cuttings can be planted on raised beds or pots.
- Holes are made and the cuttings are inserted with 1-2 nodes in the soil. Care should be taken to ensure that the slant cut only goes into the soil.
- After planting, the soil around the cutting is gently pressed so that the cutting is firm in its position.
- After planting the cutting, immediate irrigation is must.

Semi-hard wood stem cuttings

Semi hard wood cuttings are prepared from new shoots just after a flush of growth which is partially matured. Camellia, Citrus, Eranthemum, Acalypha, Geranium, Hibiscus, Jasmine, Lemon, Olive etc. are propagated by this method.

Preparation and planting

- Partially matured shoots from a healthy and vigorous growing plant are selected.
- Pencil thick slightly flexible 7 - 15 cm portion is taken by giving a sharp cut just below a basal node.
- All the leaves from the basal two third portion of the cutting are removed. Only one third terminal leaves are retained. If the plant species is with large leaves, the leaf tips should be cut and the leaf size should be reduced by half.
- The cuts should be made as in the case of hard wood cuttings.
- The cuttings are planted in the same way as hard wood cuttings.

Soft wood cuttings

Cuttings are prepared from the soft succulent new spring growth of species which are 4 to 6 months old. Nerium, Crotons, Eranthemum, Graptophyllum etc. are propagated by this method.

Preparation and planting

- A pencil thick soft succulent shoot from a healthy and vigorous growing plant is selected.
- A cutting which is 7 to 15cm long is made after giving sharp cuts.
- The basal one third leaves are removed from the cutting and leaves are retained in apical two third portion of the cutting.
- Soft wood cuttings should be kept in green house or in a mist chamber where high humidity can be maintained in order to keep the tissues in turgid condition.
- The cuttings are planted in the same way as hard wood cuttings.

Herbaceous stem cuttings

This type of cuttings is taken from succulent herbaceous green house plants. Chrysanthemum, coleus, carnations, dahlia, geraniums, cactus etc., are propagated by this method.

Preparation and planting

- Succulent herbaceous shoots from a healthy and vigorous plant is selected.
- Apical portion of the shoot, about 15 cm is selected for this purpose.
- All the leaves are retained.
- A cut is given below a basal node.
- The cuttings are planted in the same way as hard wood cuttings.

ii) **Leaf Cuttings**

Certain plants with thick and fleshy leaves have the capacity to produce plantlets on their leaves. In leaf cuttings, the leaf blade with or without petiole and axillary bud is used for starting new plants. Adventitious roots and shoots form at the base of the leaf and a new plant is formed. However, the original leaf does not become a part of the new plant. Frequent watering and high humidity and bottom heating are desirable for better and rapid rooting of leaf cuttings. Sand or sand and peat moss (1:1) are satisfactory rooting media for leaf cuttings. Leaf cuttings are of the following types -

- Leaf blade / Leaf section cutting

- Leaf vein cutting / Leaf slashing
- Leaf margin cutting
- Leaf bud cutting

Leaf blade/Leaf section cutting: Done in large leaved plants such as Sansevieria.

Preparation and planting

- A large healthy leaf is selected
- A slant cut is given towards the base of the leaf and the top cut is straight, which is given at about 10 cm from base cut.
- Similar cuts are given along the leaf till the leaf tip. Taking care of leaf polarity.
- ¾ of the prepared leaf cuttings is inserted in the medium.
- The soil around the cutting is compressed and the cuttings are watered.

Leaf Section cutting—Sansevieria.

Leaf vein cutting/Leaf slashing: The whole leaf is used to get new plantlets. *Begonia rex 0*is successfully propagated by this method.

Preparation and planting

- A healthy and full mature leaf is detached from the mother plant.
- Cuts are given on alternate veins closer to the petiole on the lower surface of the leaf.
- The leaf is laid flat on the medium in such a way that the lower portion comes in contact with the medium and the leaf should be stable in a place.
- The cuttings should be watered carefully.

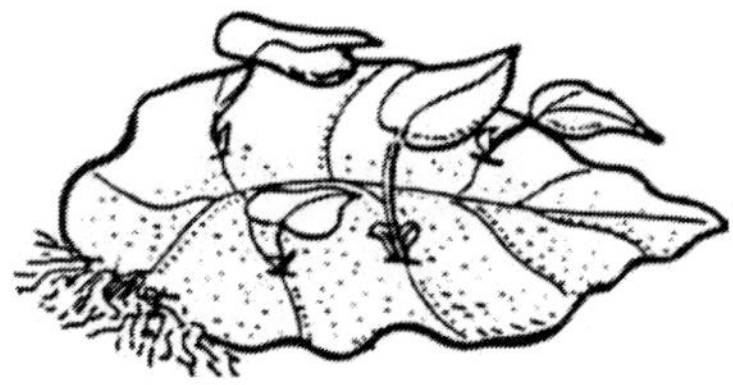

Leaf vein Cutting

Leaf bud cuttings: This cutting consists of a leaf blade, petiole and a short piece of the stem with attached axillary bud. This is practiced in species that are able to initiate roots but not shoots from the detached leaves. In such cases the axillary bud at the base of the petiole provides nutrients for the essential shoot formation E.g. black berry, camellia, lemon, rhododendron and raspberry etc.

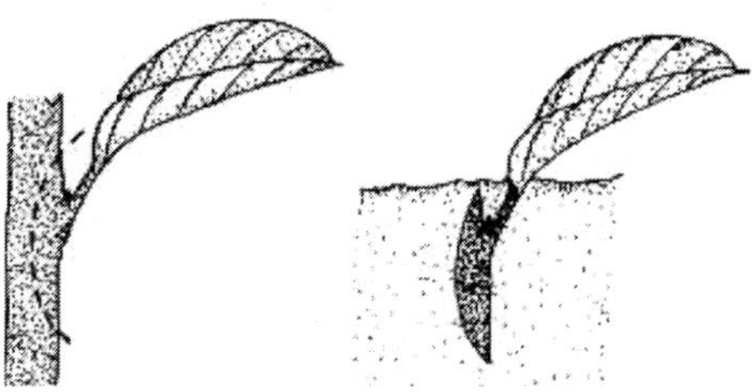

Leaf bud cutting

Preparation and planting

- A healthy and mature shoot with well-developed buds and healthy active growing leaves is selected.
- A leaf along with the axillary bud and a small portion of the stem is separated.
- If necessary, the cut surface of the prepared cuttings may be treated with the recommended root promoting substance to stimulate rapid root formation.
- The prepared cuttings are inserted in the rooting medium so that the bud is1.5 to 2.5 cm below the surface.
- The medium around the cutting is compressed and immediately irrigated.

Leaf margin cutting: This technique is used to propagate plant material like Bryophyllum.

Preparation and planting

- A mature and healthy leaf with intact foliar embryos is selected.
- The leaf is kept flat on the medium and partially covered with soil in order to keep the leaf margin in contact with the medium.

- The cuttings are carefully watered.

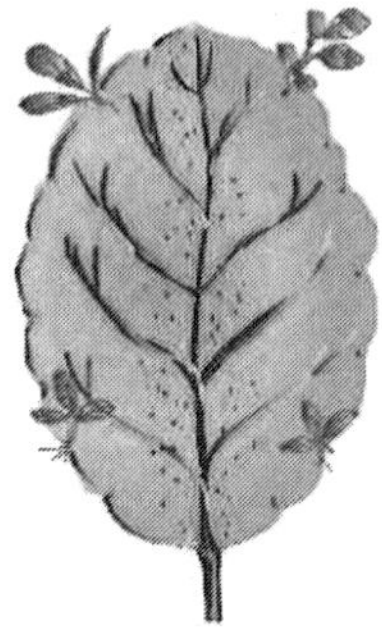

Leaf margin cutting of Bryophyllum

iii) **Root cuttings:** Plants which give rise to root suckers freely are propagated by root cuttings. Eg. bread fruit (*Artocarpus altilis*).

Root cuttings

B. Layering

Layering is a technique where the roots are developed on a stem while it is still attached to the parent plant. The rooted stem portion is detached from the mother plant and it becomes a new plant growing on the newly initiated roots. A layered stem is known as a 'layer'. Layering includes several forms of ground and aerial layering. When rooting is encouraged in a plant part which does not touch the ground, it is known as 'air layering' or 'gootee' or 'marcottage'. When branches running parallel to ground are utilized, it is known as ground layering. The root formation during layering on a stem is stimulated by various stem treatments like ringing, notching etc., which cause an interruption in the downward translocation of carbohydrates and other growth factors from leaves and growing shoot tips. However, the root formation in layered stems completely depends upon continuous moisture supply, good aeration and moderate temperature around the rooting zone. Sometimes synthetic growth regulators like IBA, IAA etc, are also applied to the layered stem, to induce better rooting, as the auxins in layered stem is an important factor for rooting.

Advantages

- It is an easy method and does not require much care and arrangement like cuttings.
- The mother plant supplies nutrient and other metabolites as it remains attached while rooting.
- By using a large branch, a number of plants can be obtained.
- True to type plants are produced.
- Some plants that cannot be satisfactorily propagated by other methods,can be propagated by layering.

Disadvantages

- It is a slow process.
- Less number of plants as compared to cuttings can be propagated.
- Layered plants are generally shallow rooted.
- Layers interfere with cultural practices of the mother plant.
- The beneficial effect of root stock cannot be exploited.

Types of layering

i) Ground layering

Ground layering is when the layered portion, the portion where we expect roots, is in contact with the ground. It is of different types as follows -

- Tip layering
- Simple layering
- Trench layering
- Mound layering or stool layering
- Compound or serpentine layering

Tip layering

It is generally done in plants which have trailing type of shoots. Most of the berries like purple and black raspberries, and trailing blackberries are propagated by this method.

Procedure

- A hole 3 to 4 inches deep is dug near the tip of the trailing plant.
- The tip of a current season's shoot is inserted in the hole and covered with soil.

- The tip grows downward first, then bends sharply and grows upward.
- Roots are formed at the bend and the re-curved tip becomes a new plant.
- The tip layer is removed and planted.

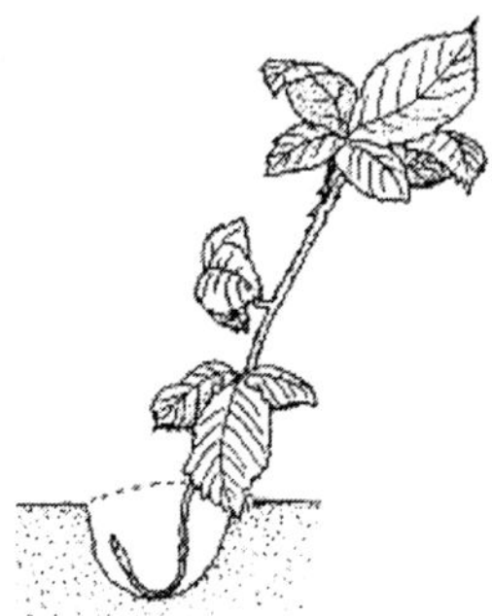

Tip layering

Simple layering

In this method, a branch is bent to the ground and some portion of it is covered by soil leaving the terminal end of the branch exposed. Root initiation takes place at the bent and buried portion. After allowing sufficient time for root formation, the rooted stem is separated from the mother plant. Eg. Bougainvillea, Jasmine, Rangoon creeper.

Procedure

- A healthy, flexible and sufficiently long (50 to 60 cm) branch is selected.
- The selected branch should be closer to the ground.
- At a distance of about 15 to 30 cm back from the tip, a sharp, slanting inward and upward cut is given, 1.5 to 2.5 cm below a node and a small wood splinter is inserted.
- The shoot is gently bent to the ground so that the treated part is conveniently inserted into the soil.
- The treated region is covered with soil.
- The shoot is pegged down or a stone or brick is placed on the soil covered area, to keep the layered shoot in place.
- The terminal portion is kept upright by tying to a stake.
- The layered portion should be watered regularly since moisture is important for root initiation.
- After sufficient root formation, the layer is separated by cutting just below the rooted zone.

Simple layering

Simple layering or ground layering

Compound or serpentine Layering

Compound layering is essentially the same as the simple layering except that the branch is alternatively covered and exposed along its length. The branch for compound layering must be long and flexible so that it can be layered at different places along its length. Eg. bougainvillea, jasmine, rangoon creeper.

Procedure

- A healthy, flexible and sufficiently long (100 to 250 cm) basal branch that is close to the ground is selected.
- A sharp slanting, inward and upward cut, 1.5 to 2.5 cm below a node is made at 30 cm interval starting from the tip leaving, 2 to 3 buds in between two such cuts.
- The shoot gently bent and inserted in the ground.
- The cut portions are covered with the soil and the uncut portions are exposed.
- The remaining steps are same as in simple layering.

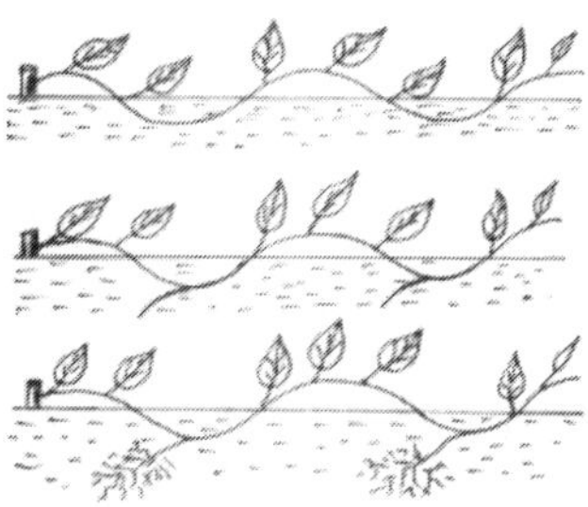

Compound or serpentine layering

Mound (stool) layering

In this method, a plant is cut back to the ground level during the dormant season and when the new shoots grow, soil is heaped around the base. After allowing sufficient time for root initiation, individual rooted layers are separated from the mother plant and planted. E.g. apple roots stocks, guava, litchi, quince.

Procedure

- The plant to be mound layered is selected.
- The plant is cutback to 2.5 cm above the ground level, just before growth begins.
- The new shoots are allowed to develop.
- When these shoots are 7 to 15 cm tall, girdling is done at the base.
- The shoots are covered in a mound of soil to half its height.
- When these shoots are 20 to 25 cm tall, again soil is added to raise the mound to half the height of the shoot. Rising of mound is again done when the shoots are about 35 - 45 cm tall.
- The mound should be watered regularly to initiate rooting of shoots.
- A depression can be made in the centre of the heap to hold water.
- After sufficient root formation, the soil mound is carefully removed and the rooted shoots are cut and planted.
- The rooted shoots are transplanted.

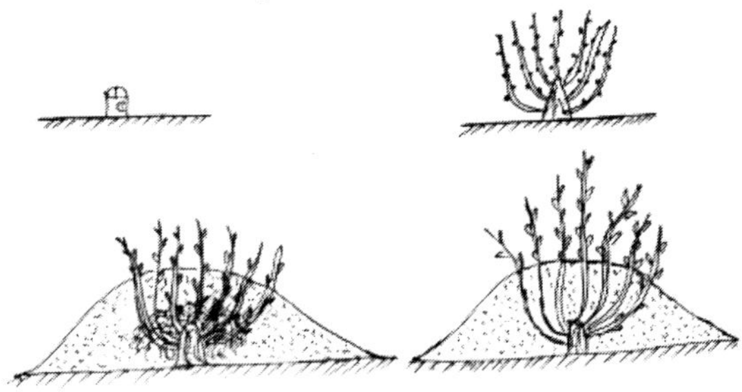

Mound or stool layering

Trench Layering

Trench layering consists of growing a plant or a branch of a plant in a horizontal position in the base of a trench and filling in soil around the new shoots as they develop, so that the shoot bases are etiolated. Roots develop from the base of these new shoots. Trench layering is used primarily for woody species which are difficult to propagate by mound layering. Eg. apple rootstocks, litchi, quince.

Procedure

- Small trenches of about 25-30 cm deep are dug in 1 m wide rows.
- One year-old plants are planted in the trenches in rows at an angle of 30^0 to 45^0.
- The plants are spaced 5 to 10 cm apart within the row.
- The rows should be 1.2 to 1.5 m apart.

- Just before growth is about to begin, the plant or a branch is laid flat on the bottom of the trench, positioned with wooden pegs.
- After some time, the buds on the branches swell and put out upright shoots.
- The shoots are slightly cut back and the weak branches are removed.
- The base of the shoots is layered and rooting medium is applied.
- The new rooted shoots are cut and planted.

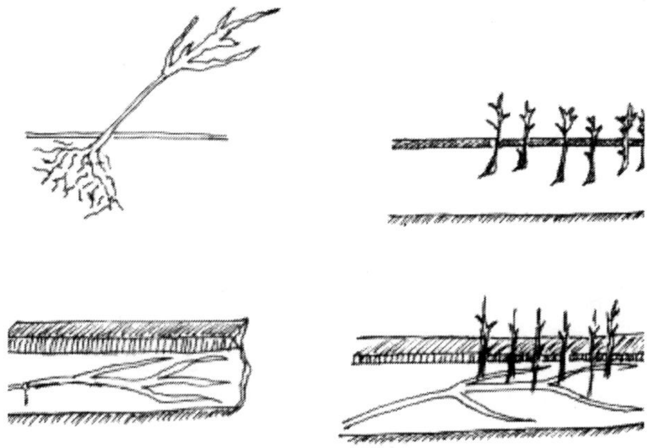

Trench layering

ii) Air layering

In air layering roots form on an aerial shoot. The rooting medium is tied to the shoot for root initiation. Sphagnum moss is the best rooting medium for air layering as it holds large quantities of water till root initiation. Eg. crotons, ficus, fig, guava, phalsa, pomegranate.

Procedure

- A healthy branch of previous season's growth is selected.
- A 2.5 cm circular bark is removed from the shoot to make a girdle.
- The girdle should be made near a healthy bud, 25-30 cm away from the tip of the shoot.
- The girdle should be free from traces of bark; otherwise the plant will heal by producing the bark, instead of roots.
- Handful of moss is placed on the girdle and it is tied with polythene.
- After a few days roots are visible through the polythene.
- The air layer is separated as a new plant and planted.

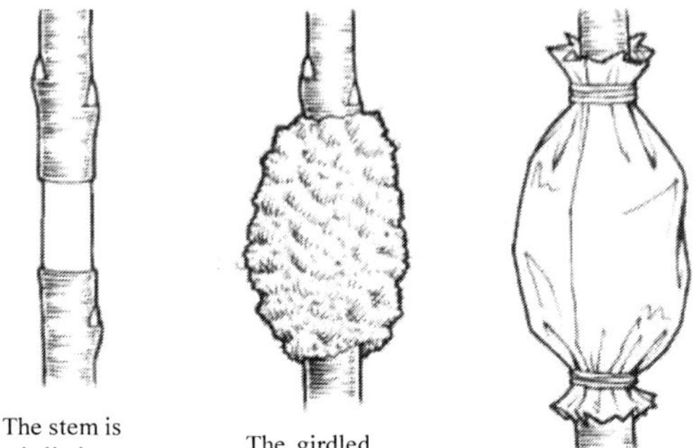

The stem is girdled to induce root formation abobve the cut

The girdled stem is covered with damp moss

Aluminium foil or plastic around the moss and tied at both ends. This cover is removed 2-3 months after tying or when the roots can be seen

Air layering

Steps in air layering

Removal of bark

Placing media for rooting

Air layered plant

C. Grafting

Grafting is an art of joining parts of two independent plants in such a manner that they unite and grow together into single independent plant. This procedure is mostly done in fruit crops. The part of graft combination which is to become the upper portion or the shoot system or top of the new plant is termed the **scion** or **cion** and the part which is to become the lower portion or the root system is the **rootstock** or **under stock** or **stock**. The scion is selected on the basis of the quality of its produce and a rootstock is sturdy and stress tolerant, even if it is not a quality producer. None of the branches of the rootstock are allowed to proliferate; it forms only the root system. In a nursery the rootstock is produced and the scion is collected from various sources to graft on the rootstock to produce a desirable graft union. The rootstock production is done as follows:

Root stock production from seeds

- The seeds of the desired plant are collected from a good source like local market or processing units (e.g, mango, sapota, aonla). The seeds should be immersed in water, only the sinkers are selected and floaters are discarded.

- As for all the nurseries, the rootstock nursery should be rich in farm yard manure and should be in a semi-shade area.
- The collected seeds are sown and raised directly in polybags or nursery beds at an appropriate spacing.
- If sown in nursery beds, the seedlings should be gently lifted after a month or so, and planted in polybags or earthen pots, so that grafting scion is easy.
- Rootstocks are also raised by cuttings, as in case of rose, cactus etc.

Methods of grafting

Mainly in grafting there are two types *viz*.,

- Attached scion methods of grafting - Simple inarching / Approach grafting
- Detached scion methods of grafting :
    - o Side grafting – and e.g., grafting Veneer
    - o Apical grafting, Epicotyl grafting, Softwood grafting, Cleft grafting

In attached scion methods of grafting the scion remains attached to the mother plant (scion plant) till the graft union takes place. Hence it is important that the chosen scion mother plant is nearby or in the nursery itself as rootstock has to be taken to the vicinity of the scion. In this method of grafting simple inarching or approach grafting is most important.

Simple inarching / Approach grafting

The distinguishing feature of this method of grafting is that two independent plants which are on their own roots (self-sustaining) are grafted together. This method provides a means of establishing a successful union between certain plants which are difficult to graft by any other method as the two plants will be on their own roots till the formation of successful graft. In this system, even if the graft union is unsuccessful, the rootstock and scion, both are not lost. E.g. mango, sapota.

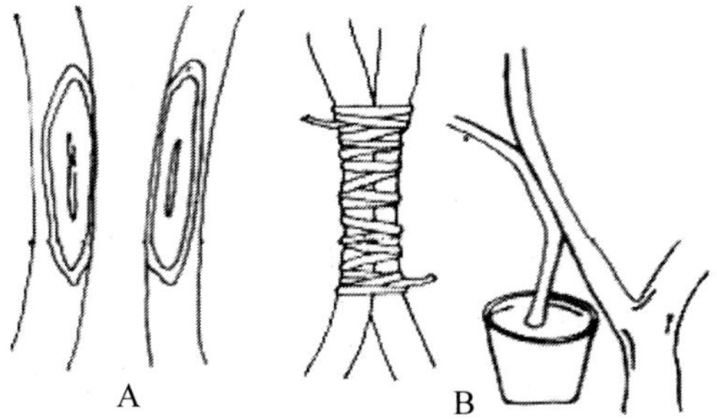

Simple inarching

Procedure

- A healthy shoot of 3.5cm girth is selected on the mother plant, which is the source of scion.
- A root stock (raised in pot) is selected. This should have approximately the same stem thickness as that of the selected scion shoot on the mother plant.
- On the internodal region, a slice of 2.5 to 5 cm long bark and wood is cut from both the selected stock and scion shoots.
- The cut should be given on the stock and scion should be of the same size.
- The cuts should be perfectly smooth so that when the stock and scion are pressed together, a close contact of the cambial layers occurs.
- The two cut surfaces facing each other, are tied tightly together with string or cloth.

Approach grafting in sapota

Separation of grafts from mother plant

Detached scion method of grafting

In the detached scion methods of grafting the scion is separated from the scion plant or mother plant just before grafting. A partially matured pencil thick scion shoot is selected in the mother plant. The selected shoot is defoliated retaining only the petioles up to a length of about 10 cm from the apical bud. The defoliated shoot is left on the tree for a period of 7-10 days. During this time, the bud on the shoot begins to swell. This shoot is then called as **pre-cured scion,** and the process is called pre-curing.

In detached scion methods of grafting there are two types they are –

Side grafting and Apical grafting

Veneer grafting is a kind of side grafting. It is used widely for grafting small potted plants and *in situ* grafting. E.g. avocado, mango etc.

Procedure

- On the stock plant, at the desired height, in the internodal region, a shallow inward cut running to a length of about 2.5 to 5 cm is given.

- At the base of the first cut, another short and inward cut intersecting the first cut is given and a piece of wood and bark is removed.
- On the scion, towards the base, a 2.5-5 cm long, slanting cut towards one side and another short, inward cut is given on the opposite side.
- The cuts given on stock and scion should be of same dimensions, so that, the cambial layers can be matched as closely as possible.
- The scion is inserted on to the rootstock such that a contact of cambium is established and tied firmly.
- After the union has healed, the stock is cut back above the graft union either gradually or all at once.

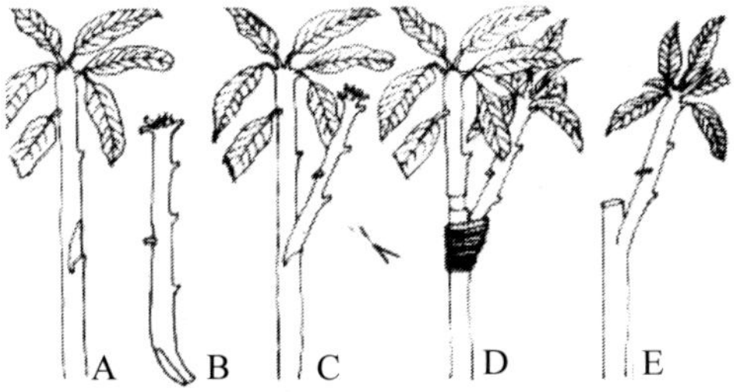

A-Prepared root stock, **B**-prepared scion, C- Scion inserted, **D**-Girdled stock and tied graft joint, **E**- Successful graft, the stock being removed

Apical grafting

It is done on the apical region of the rootstock seedling. Apical grafting is of the following types.

Epicotyl / Stone grafting

This method of grafting is done on the epicotyl region of the young seedlings; hence the name epicotyl grafting. **Eg.** cashew, mango etc.

Procedure

- The top portion of the chosen rootstock seedling is cut - off, leaving 5-6 cm long shoot (epicotyl).
- With a sharp knife a 2-3 cm long vertical, downward slit is made at the centre of the shoot.
- For scion, a 3-4 months old, 5-8 cm long, dormant terminal shoot is selected.
- The lower end of the selected scion is cut to a wedge shape by giving slanting and inward cuts of 2-3 cm on the opposite sides.
- The wedge shaped scion is inserted in the slit made on the rootstock seedling and secured firmly with polythene strips or tape.

- The graft is irrigated regularly without wetting the graft region.
- In about three weeks the scion starts sprouting.

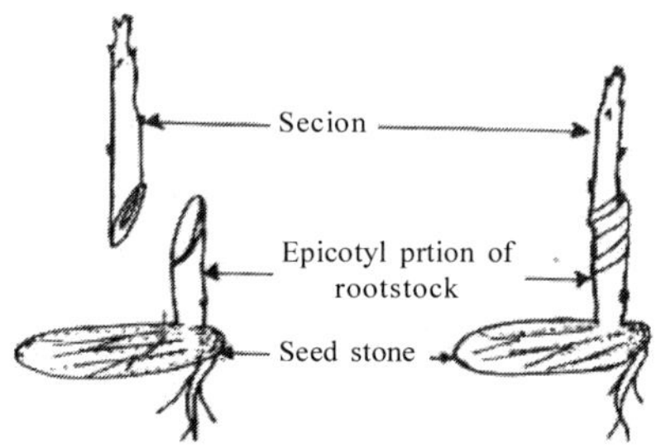

Epicotyl grafting

Soft wood grafting

This technique has been developed to graft small and young rootstocks which are grown *in situ* or in pots. Eg. cashew, mango.

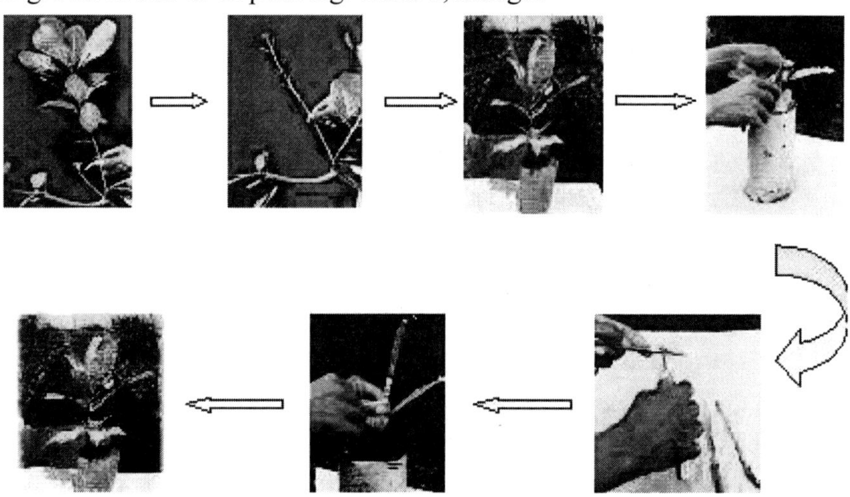

Softwood grafting

Procedure

- The rootstock seedlings are raised in suitable containers or preferably in the main field itself where the grafts are desired to be grown and allowed to grow for a year or more.
- When the seedling attains a height of 30-45 cm, the shoot is decapitated after retaining 8 cm stem.
- In the retained fresh stem, a longitudinal cut of 3 cm is made.
- A 10 cm long scion shoot, about the same thickness of the prepared rootstock seedling is selected.

- The basal end of the scion is cut into a 3 cm wedge by chopping the bark and a little wood on two opposite sides.
- The prepared wedge part of the scion shoot is inserted into the slit made on the rootstock and secured firmly with polythene strips.
- The grafted plant is irrigated regularly and scion sprouts in about three weeks.

Cleft grafting

Cleft grafting technique is used for top working, to change a variety on healthy long lived plant species. This may be done in a strong, healthy, old tree, which does not produce quality material or has become unproductive, so the variety is changed by top working by cleft grafting, while retaining the root system. Short lived species, old trees or diseased trees are not suitable for top working; in such cases new planting is considered more economical and useful than top working.

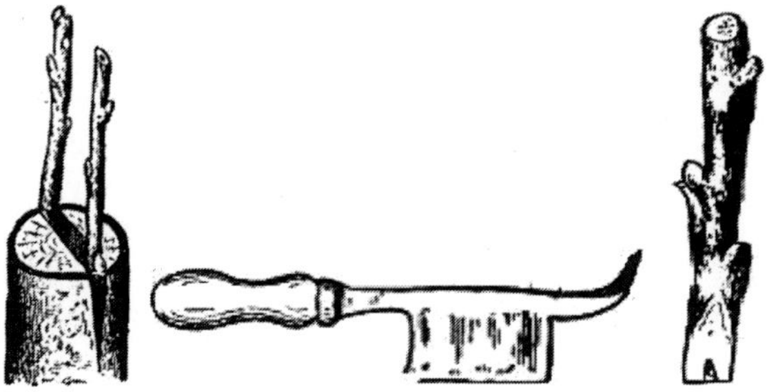

Top working by cleft grafting

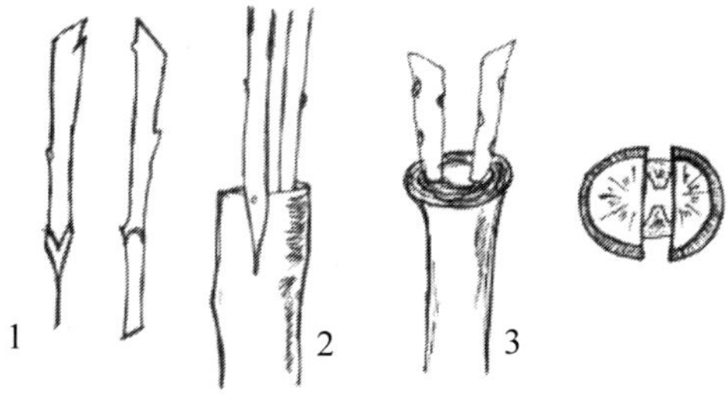

Top working is done when

- When the existing tree is of inferior type.
- When the tree is unproductive.
- To provide pollenizers.
- To change the variety.

For top working different methods of grafting like cleft grafting, bark grafting, splice grafting or side grafting can be used. However, cleft is the most popular and commonly used method especially when there are old thick branches. When young and thin branches are present, whip and tongue grafting are the best.

Top working of older trees is generally done over a period of two years. In the first year, half of the scaffold branches are top worked retaining the other branches as nurse branches which in turn are grafted in the second year. In the smaller and comparatively younger trees the entire tree is top worked in the first year. Here also one or more nurse branches are retained till the union is successful. Nurse branches protect the top worked scions from winter injury, sun burn and also from desiccating winds and development of water sprouts. Top working is most successful when relatively young trees are used. If older trees are selected it is better to select vigorous lateral branches that arise from the main branch. The branches to be top worked should be given a clean and smooth cut. The cut should be at a point where there are no small branches or knots on the branches. The selected scion shoots should be cleanly cut. Immediately after top working the limbs should be thoroughly covered with grafting wax, sealing all the exposed cut surfaces.

Procedure

- The old tree is treated as a rootstock.
- The branch of the old tree is split cleanly.
- The scion is prepared by making a long tapering wedge.
- The split in the rootstock is gently opened and two scions are carefully inserted on either side of the rootstock split.
- Grafting wax is filled.

Stock-scion relationship

After grafting, specific interactions between scion and root stock may result in alterations on size, growth, productivity, fruit quality or resistance to pests and diseases in the grafted plant. These changes may be either from scion or from root stock.

Effects of root stock on scion

i) **Size and growth habit:** Size control is one of the most significant effects of rootstock on scion. E.g. dwarf, semi-dwarf, vigourous and very vigourous rootstocks are available in apple. Root stock effects on tree size and vigour are noticed in citrus, pear and other species.

ii) **Precocity in fruit set and yield:** Fruitfulness of a tree can be influenced by the rootstock. In general, fruiting precocity is associated with dwarfing rootstocks and slowness in fruiting is associated with vigorous rootstocks. E.g. In grapes, the bearing occurs in current season's growth and the rootstock are strong influencing factor. High yield obtained in certain american type of grapes (*Vitis labrusca*) when grafted on vigorous rootstocks when compared to its own rootstock. On the other hand, in *Vitis vinefera* scions, use of "Dog Ridge" (*Vitis champini*), which is an extremely vigourous rootstock, produce strong vines in fertile soils, finally become unproductive also.

iii) **Size, quality and maturity of fruit:** Considerable variation was noticed among varieties in respect of size, quality and maturity when different root stocks were used. E.g. Fruit size of both "Washington Navel" and "Valencia"orange is strongly influenced by root stock. The largest Navel Orange fruits are produced on Sour orange rootstocks and the smallest on the Palestine sweet lime. Similarly, the largest "Valencia"oranges are produced on dwarfing trifoliate orange stock, and the smallest fruits on sweet orange root stocks.

iv) **Stress tolerance**: Rootstock shows profound effect on biotic and abiotic stresses like pests and disease resistance, drought tolerance, water logging conditions, cold hardiness etc. E.g. In grapes, "Dog ridge" stock can be used to impart resistance to nematodes, while "Salt creek" for salinity tolerance. In mango, Israel 13-1 can be used as salt tolerant root stock. In sapota, the plants grafted on Pala root stock (*Manilkara hexandra)* performs well under both drought and water logging conditions than on own root stock. In citrus, the rootstock can impart cold hardiness on the scion cultivars. Grape fruit trees grafted on rangpur lime survived much better under severe cold conditions than those on rough lemon or sour orange.

Effects of scion on root stock

Although the effect of rootstock on scion is more prominently noticed in grafted plants, in certain combinations the scion had also altered the characters of root stock.

i) **Improvement in vigour of rootstock:** This is the major influence of the scion on the stock. If a strongly growing scion cultivar is grafted on a weak root stock, the growth of the rootstock will be stimulated. Different scion cultivars may cause a characteristic root growth pattern in the root stocks. E.g. In apple, the root stocks "Malling 9" and "Malling 12" when grafted, marked influence was noticed on the quantity of roots produced with different scion cultivars.

ii) **Cold hardiness of the root stock:** Cold hardiness of citrus roots is affected by the scion cultivar. Sour orange budded on to "Eureka lemon" affected by winter injury than unbudded seedlings.

D. Budding

Budding is also a method of grafting wherein only one bud with a piece of bark only, sometimes with wood is used as the scion material. It is also called as bud grafting. The plant that grows after union of the stock and bud scion is known as budded plant.

Methods of budding

There are a number of methods of budding. Few important ones are explained below.

T-budding / Shield budding

This method is known as T-budding as the cut given on the rootstock is of the shape of the letter T. It is also called as shield budding since the bud piece is like a shield. This method is widely used for propagating fruit trees and many ornamental plants. This method of propagation is easily done when the stock is 0.75 to 2.50 cm in diameter and actively growing so that the bark readily separates from the wood examples citrus rose etc. For a successful 'T' budding, the scion material should be mature, dormant buds and the rootstock should be in. The bark should easily peel without damage. This happens when the vascular cambium is actively growing.

Procedure

- A root stock is selected. At 15-25 cm from ground level, a 2.5 - 3.5 cm long vertical cut is given.
- At the top of this vertical cut, another horizontal cut (1cm or $1/3^{rd}$ of the circumference of the stem) is given in such a way that the two together make a 'T'.
- To remove a bud from the selected scion shoot, the bark is sliced 1.5 cm below the bud to 2.5 cm above the bud.
- Another horizontal cut is given about 1cm above the bud to resize the bud.
- The shield of bark from behind the bud piece is gently removed, without causing a hole in the bud. All traces of the bark from bud is removed.

- The bud is inserted between the flaps the already made 'T' on the bark of rootstock in such a way that the horizontal cut of the scion bud shield matches the horizontal cut on the stock.
- The budded area is wrapped with a polythene strip exposing only the bud.

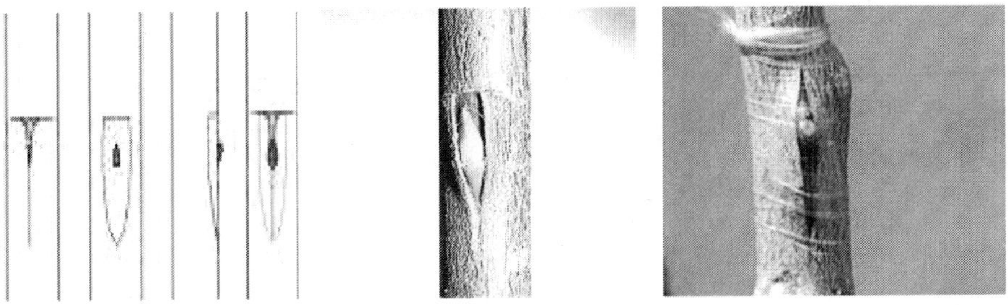

T- budding / Shield budding

Inverted T budding

In heavy rainfall areas, water running down the stem of the stock may enter the 'T' cut, soak the bud under the bark and thereby prevent healing of the bud piece. Under such conditions an inverted T budding may give better results as it is more likely to shed excess water. "Inverted T budding" procedure is same as that of T budding except the horizontal cut on the root stock is made at the bottom of the vertical cut rather than at the top.

Procedure

- On the selected rootstock plant, a horizontal cut is given at the bottom followed by a vertical cut representing inverted T.
- The required bud shoot is selected and the bark is sliced 1.5 cm above the bud and 2.5 cm below the bud.
- Another horizontal cut is given about 1cm below the bud and the bud piece is removed.
- The bud is inserted between the flaps of bark on the rootstock and pushed upwards till the horizontal cut of the shield matches the horizontal cut on the stock.
- The budded rootstock is completely wrapped tightly with polythene strip, taking care not to cover the bud.

Patch budding

In this method a regular patch of bark is completely removed from the rootstock shoot and is replaced with a similar sized patch scion bark with a bud. For this method to be successful, the bark of the rootstock and scion bud shoot should be easily slipping.The diameter of the rootstock and scion bud shoot should be preferably same (1.5 to 2.75cm) Eg. ber, citrus, cocoa, rubber.

Procedure

- On the selected rootstock plant, at 10-15 cm above ground level, two horizontal parallel cuts are made on the bark.
- The cuts are 1-1.5 cm long or 1/3rd of the rootstock girth and 2- 3 cm apart.
- The horizontal lines are joined with two vertical lines, to make a rectangular patch.

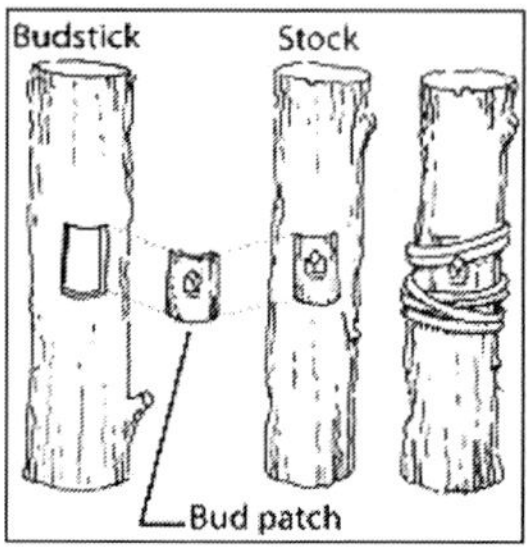

Patch budding

- The patch of bark is removed and a similar patch from the scion shoot is attached here.
- The patch with scion bud is tied tightly, exposing the bud alone.

Ring budding

The bud is prepared by taking a 3 cm ring of a bark, with the bud in the centre. In the root stock, two transverse cuts, 1.5 cm apart are made and these are connected with a vertical cut and a ring of bark is removed. The prepared scion bud with the ring of bark is fitted in the exposed portion of the rootstock and tied. Eg. cinchona.

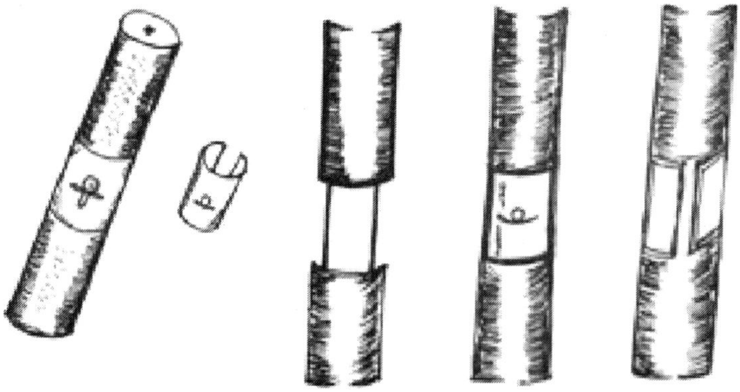

Ring budding

Double working

The process of double working has an interstock between the rootstock and the scion. Top working of grafted orchard trees is essentially a double working; here the tree trunk as an intermediate stock may exert certain influences on the new top. The inserted intermediate stem piece is called as sinking scion / foster mother /interstock / inter stem. It is practiced for several reasons:

- To overcome incompatibility between the stock and scion. Incompatible stock and scion may be united by means of a piece of interstock that is compatible to both.
- To secure resistance to drought or cold by providing a resistant trunk by means of double working.
- To obtain resistance to pest and dwarfing effect by using a pest resistant stock and a dwarfing interstock.

E. Propagation through specialized plant parts

Many annual plant species die the end of the growing season and come back again the next year. These plants have underground food storage organs that survive the dormant period. These underground food storage organs are also used for vegetative propagation. These when planted, give rise to new plants. Some of the underground structures can be separated easily and many new plants can be made, while some structures have many sprouts coming out and each sprout can be cut out and planted to get a new plant. Hence the ground storage organs can be grouped into two classes based on how they are propagated viz., plants propagated by separation and plants propagated by division.

Plants propagated by separation

Separation is a method of propagation in which underground structures of plants are divided not by cutting but by breaking along natural lines between segments. Separation is breaking away of daughter structures from the mother structure to be used to establish new plants. Specialized underground structures *viz.*, bulbs and corms produce such material.

Bulb

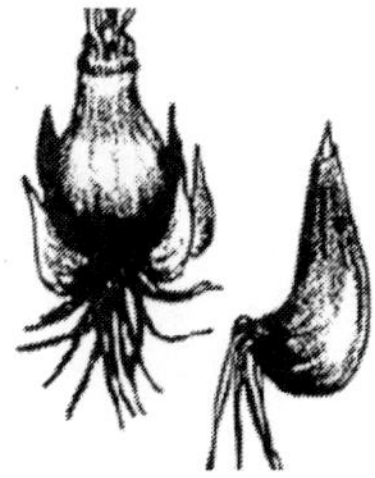

Separation of bulblets

Bulb: A bulb is a specialized underground organ that consists of fleshy leaf scales growing on a flat stem basal plate. The scales wrap around a growing point tightly to form a ball like structure. Small bulbs originate in the axils of some of these scales and these can easily be, to be used as a new bulb which would give rise to a new plant.

There are two types of bulbs- ***tunicate*** and ***non-tunicate bulbs***.

Tunicate-These bulbs have outer bulb scales that are dry and like membrane. This covering called tunic, provide protection from drying and mechanical injury to the bulb. The fleshy scales are in continuous, concentric layers, called lamina.E.g. onion, daffodil, tulip etc.

Non-tunicate (scaly) bulbs: These bulbs do not have the dry covering. The scales are separate and attached to the basal plate. The scales are loose and can be separated individually from the bulb. The non-tunicate bulbs are easily damaged and must be handled carefully. The daughter bulbs or bulblets develop at the base of the scales of the mother bulb. Eg. lily.

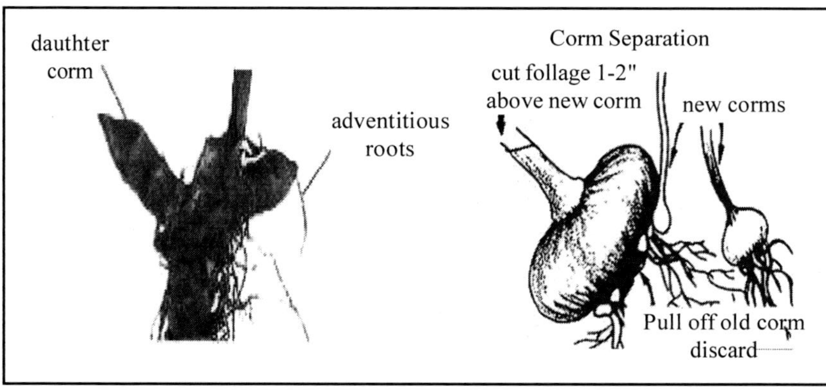

Corm: The corm is a modified stem with modified leaves. Food is stored in compact stem, which has nodes and very short internodes and is wrapped up in dry, scaly leaves. When a corm sprouts into a new shoot, the old corm becomes exhausted of its stored food and is destroyed as a new corm forms above it. Several small corms, or cormels, arise at the base of the new corm. The cormels may be separated from the mother corm at maturity and used to propagate new plants. Eg. *Amorphophallus*, colocasia, gladiolus etc.

Plant propagation by division: It is a method of propagating plants using cut section of a particular food storage organ like rhizome, tuber and tuberous root etc.

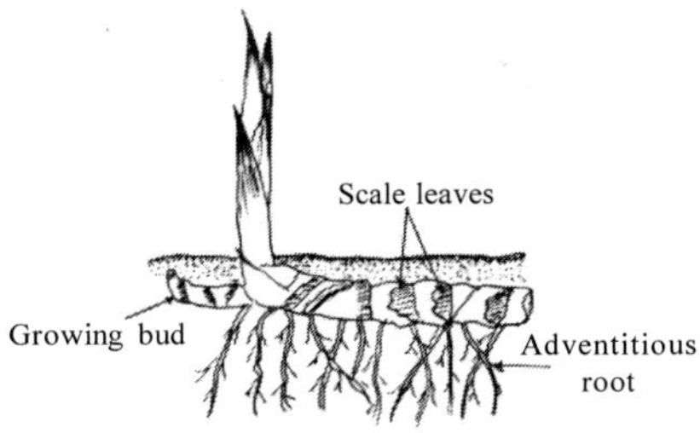

Rhizome of Canna

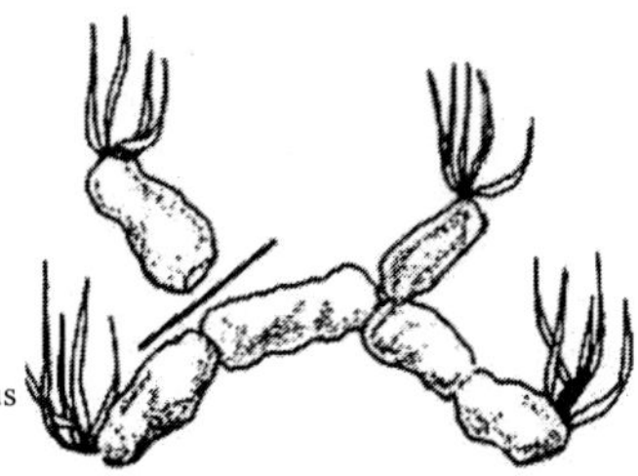

Rhizome division

Rhizome: Itis a modified stem structure in which the main axis of the plant grows horizontally just below the surface of the ground, and has roots on the underside. The stem has nodes and internodes. The rhizome is a branched clump made up of short individual sections. While propagating plants by rhizome, it should be divided in such a way that each section has at least one lateral bud or eye. Eg. bamboo, banana, iris etc.

Stolon: It is also a horizontally growing stem, which produces adventitious roots when it comes in contact with soil. These make a network of stems above the soil surface

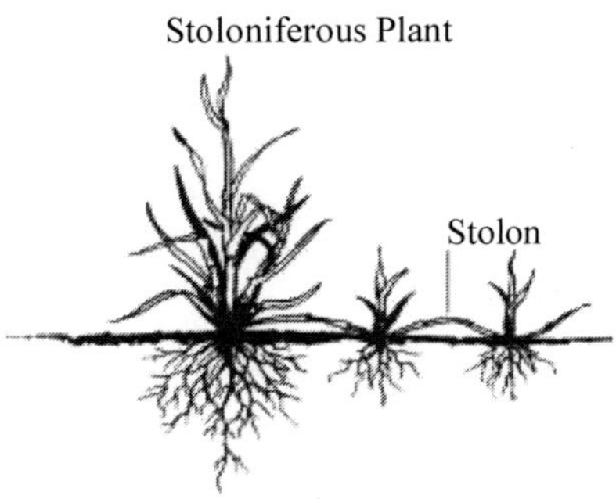

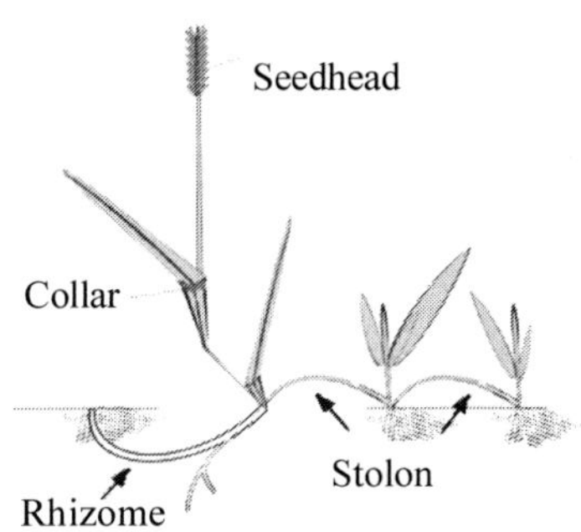

Stolon

Stolon actually is a naturally occurring layer. The stolons can be cut into small rooted segments and used as propagating material. Eg. mint, *Cynadon dactylon,* bermuda grass etc.

Runner of Strawberry

Runner: A runner is a specialized stem that develops from the axil of a leaf at the plant crown. It grows horizontally along the ground and forms a new plant at one of the nodes. The rooted daughter plants are used for propagation. Eg. straw berry, oxalis, blue berry etc.

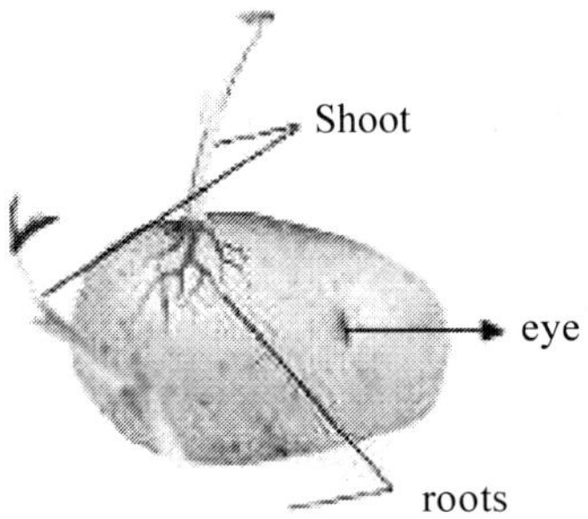

Tuber Division

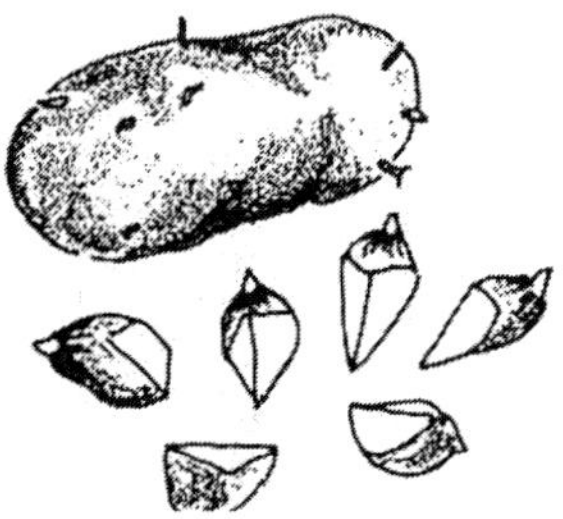

Stem tuber of Potato

Stem tuber: A tuber is also a modified stem. It is swollen underground structure having nodes on the surface, known as 'eyes'. The tuber is divided into sections so that each section has a good amount of stored food and a bud or 'eye'. Eg. potato.

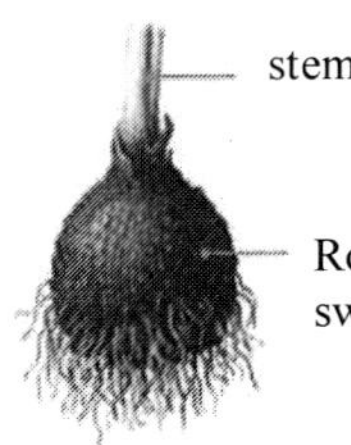

Root Tuber of Dahlia

Tuberous roots of sweet potato

Tuberous roots: These are thickened tuberous growth which has stored food. These lack nodes and internodes. Tuberous roots are propagated by dividing the crown in such a way that each part has a shoot bud. Eg. dahlia, begonia, sweet potato.

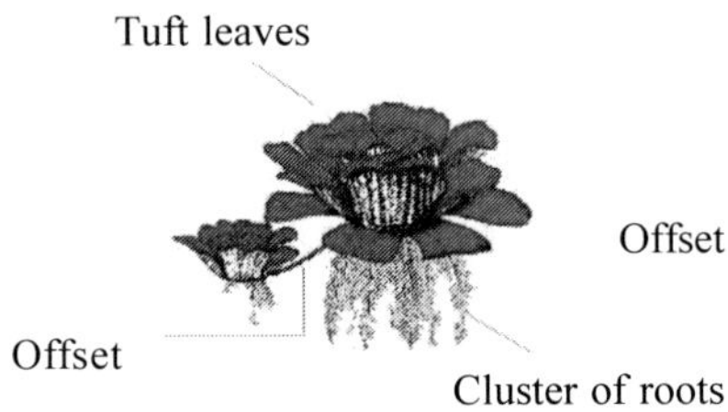

Offset of Pistia

Offset: An offset is a short thick horizontal branch growing out of the crown of a plant. The horizontal branch has a cluster of leaves at its end. This cluster of leaves develops roots and now it is the daughter offset, which can be used to propagate as new plant after being detached from the mother plant.Eg. pistia, agave, water hyacinth, cycas, dracaena etc.

Suckers: A sucker is a lateral branch developing from the underground parts of the stem or roots. The suckers emerge from below the soil surface. There are two types of suckers -

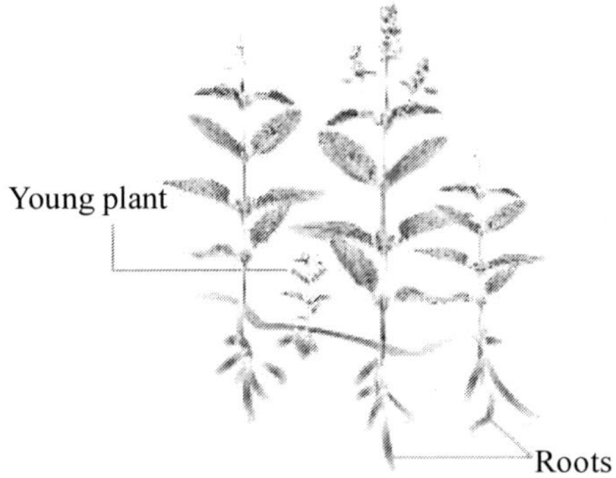

Stem sucker of Mint

a) ***Shoot suckers:*** These emerge from the base of the stem. Sometimes the suckers grow upwards and directly give raise a shoot. Mostly it grows horizontally before turning upward. Whichever way it appears; it roots while it is still attached to the mother plant. Propagation with suckers is done after separating the sucker from the mother plant. Eg. chrysanthemum, banana, pineapple, yucca.

b) ***Root suckers:*** Adventitious buds on the roots give rise to root suckers. Root suckers are separated and planted. Eg. guava, *Millingtonia*, curry leaf, *Quisquails* etc.

Slips and crown

Slips, suckers and crowns are the propagation structures for pine apple. Slips arising from the base of fruit on the fruit stalk, while crowns borne at the top. Suckers are arising from the basal stem of the plant. Sucker grown plants usually fruit earliest at around 16 months, slips may take 24 months and a crown can take up to 28 months.

Plant propagation structures

Plant propagation structures provide the environment to get the best plant growth and development. Plant growing structures are two types viz.,

- Structures which provide optimum conditions for plant growth and development - Eg. green house, hot beds, pit nursery, mist chamber and screen houses.
- Structures which are used for hardening and preparing plant propagules for planting in the field - Eg. cold frames, lath house.

Advantages of plant propagation structures

- Plants can be protected from adverse environmental conditions.
- The time of availability of the product can be altered according to market demand to get premium price in the market. Precocious and off season products attracts good price.

In India, mist chamber and pit nursery structures are commonly used by nurserymen. Large commercial nurseries utilize glasshouses for mass propagation. Hot beds, cold frames, lath houses, net houses and greenhouses are used in temperate regions where frost is severe.

Mist chamber

Mist chamber is an enclosed area covered with polyethylene sheet or fibre glass. In mist chamber sterile propagation medium is used. Systematic spray of mist at required intervals increases the relative humidity of the chamber and creates favourable conditions for formation of roots in cuttings. In the mist chamber, temperature and humidity are controlled to suit the needs of plants and type of cuttings. Simultaneously the transpiration and respiration levels also come down. This results in high regenerative ability in a wide range of plant species. Cuttings which fail to root easily can be successfully rooted in mist chambers. Besides rooting of cuttings, the epicotyl grafts of mango when kept for seven days inside the mist chamber give hundred per cent successful results.

In temperate regions, it is essential to select the site which is open to the sun whereas in tropical and subtropical regions where summer is very hot, the mist house should be installed either nearer to the building or larger trees which could provide partial shade. The building can be white washed twice a year as it helps in reflecting sun's rays to a greater extent.

Mist propagation of fruit and ornamental plants has been taken up systematically for the first time in India at the Birla Laboratory, Agricultural and Horticulture Society of India.

Guidelines for effective functioning of mist chamber

- The rooting medium should be pathogen free and well drained.
- The water used for misting should not be alkaline.
- Hygiene and cleanliness of high standard should be observed in the mist chamber.
- The misting interval and frequency should be decided on the basis of species and variety.
- The nutrients can also be applied if necessary.
- Optimum shade should be provided if the sun is too intense.
- Cross ventilation should be provided.

Mist propagation equipments

i) **Centrifuged pump with 1-2 HP motor** - The capacity of the motor should be decided based on the size of the mist chamber. Water should be allowed to get filtered before being pumped into the mist chamber.

ii) **Pressure tank-** It is a thick walled air tight metallic chamber. Water is filled automatically when pressure drops below the limit.

iii) **Timer-** It controls the duration and frequency of the mist spray.

iv) **Nozzles-** There are various types of nozzles. Deflection type of nozzle is the best suited for medium sized mist chambers.

Pit nursery

Pit nursery is highly suitable for tropical regions, like plains of Tamil Nadu, Kerala and Andhra Pradesh. High temperature (>30^{0}C) prevailing under tropical environment has been found to be the most important limiting factor in nursery production. High temperature increases the transpiration rate leading to drying and desiccation of cuttings. The callus gets dried up quickly leading to failure of graft and bud union. High temperature coupled with low humidity proves to be more hazardous than high temperature alone. Pit nursery is advantageous to overcome those two limiting factors.

In pit nursery system, a pit of 10’x 40’x 3’ is dug and filled with water and allowed to soak and drain. Repeated water sprays are given frequently to maintain high humidity and ideal temperature of less than 25ºC in the pit. The containers with cuttings are placed in the pit nursery and covered with a polythene sheet. The polythene cover prevents evaporation of water from pit. The pit should be located under shade.

The pit nursery is practically suited for rooting herbaceous cuttings and union of epicotyl grafts and soft wood grafts. Single nodal cuttings of many species can be easily made to root under pit. The outer boundaries of the pit nursery should be protected by raising bunds which prevents the entry of rain water.

Greenhouse/Glasshouse

A green house is a framed structure, which is covered with transparent or translucent material (usually stabilized LDPE film). It is large enough for a person to walk inside and carry out cultural operations. It is used to grow crops under fully or partially controlled environmental conditions.

The plastic film allows solar radiation to pass through, but traps the thermal radiation emitted by the crops inside the green house, thus contributing to the greenhouse effect. The carbon dioxide released by the plants at night is also trapped inside and this increases the carbon dioxide level available to the plants during the day time by 5-10 times, thus enhancing photosynthesis. Since the evaporation from the soils stays inside, it raises the humidity, consequently less irrigation is required. In all, the photosynthesis increase nearly 15 times, which naturally makes the plant performance better.

Commercial green houses are usually independent structures of even span roof construction. Several single glass houses are often attached side by side eliminating the cost of erecting side walls. Permanently attached rolling benches are provided inside.

Uses of a green house

i) Suitable for different soil and climate region

India has great variety of climates and agricultural practices. There are large tracts of land, which are too hostile for agriculture resulting in minimal cultivation in such regions. Green house creates favourable environment to facilitate the plant growth. Green house provides a feasible solution for raising vegetables and horticultural crops which improves crop productivity.

ii) Need around big cities

There is always a high demand for fresh vegetables and ornamental plants round the year. In cities, there exist continuous demand for fresh, off season, exotic flowers, vegetables and herbs. Green house could therefore be promoted to meet the requirements adequately.

iii) Green houses for plant propagation

Plant propagation in green house could mean a considerable reduction in the total time for propagation of saplings.

iv) Green house for cultivation of rare medicinal plants

India has a wide variety of orchids / herbs that have been identified for large scale cultivation. Greenhouses could provide the right type of environmental conditions for intensive cultivation of these plants.

Requirements of green house

- A good site can make a difference in the performance of green house.
- Green houses should be located away from other buildings and trees to avoid obstruction of sunlight.
- An east – west oriented green house is preferred as it maintains good winter light level and daily temperatures.
- Ground slope for drainage is an important factor. Adequate provision should be made to divert surface water away from the green house.
- A greenhouse needs a dependable source of energy for uninterrupted and predictable functioning of environmental control devices.
- A dependable supply of good quality water is needed for a green house.

Net Houses

A net house or shade house or shade net house is a structure which is enclosed by agro net or any other woven material. It allows required sunlight, moisture and air to pass through the gaps and creates a conducive environment for good plant growth.

Uses of Shade Net House

- Good for cultivation of high value vegetables, fruits, flowers, foliage, spices and medicinal plants.
- Helps in growing and maintaining propagules.
- Used in production of grafted saplings during hot summer days.
- Used to harden tissue culture plants.
- Helps in quality drying of various agro products.
- Used to protect crops and nurseries against pest attack.
- Provide considerable protection from wind, rain, hail, frost and high temperature.

The frame of a shade house provides support for cladding material. A frame of mild steel (ms) angle has a life of 20 to 25 years while a bamboo frame lasts for about 3 years.

The cladding material lasts for 3 to 5 years. Shade nets are available in different colours with wide range of shade percentages *viz.,* 25%, 30%, 50%, 60%, 75% and 90%.

Hot beds

Hot beds are an old style structure for providing raised temperatures to crops, when the outside temperature is too cold. These were popularly used prior to use electricity. Horse manure was pilled under each hot bed to generate heat in temperate regions. Nowadays heat is provided under the medium by electricity, hot water and steam pipes in temperate areas.

Hot bed is a large box like structure with a slopping, tight fitting top cover. It is filled with 10-15 cm of rooting or germinating medium over the heating cables.

Cold frames

The primary use of cold frames is conditioning or hardening of rooted cuttings. Cold frames construction is the same as hot bed except that no provision is made for supply heat from bottom. A close watch on ventilation, shading, watering and protection from severely low temperatures is necessary for success with cold frames.

Lath house

Lath houses provide shade and protect container grown plants from extremely high temperatures and light. Woven plastic material is used to provide shade. By providing shade, the lath house reduces the moisture stress and decrease the water requirement of plants.

11

Nursery Techniques

Pot mixture is the media, used in the pots for propagation of nursery plants. It is the combination of different materials in a particular proportion to create better environment for rooting, growth and development of plants.

The most commonly used ingredients to make potting mixture areas follows

- Soil
- Sand
- Coir pith
- Leaf mould
- Sphagnum moss
- Farm Yard Manure
- Compost

Some other, not so common ones are peat, vermiculite, perlite, pumice, synthetic plastic aggregates, etc.

Soil

Different kinds of soils occur in different areas. For instance, red, black, loam, clay, sandy etc. Soil forms an integral part of potting mixture.

Sand

Sand is the heaviest of all rooting media. It is chiefly silica complex with no minerals and buffering capacity. It is used alone in the potting mixture when rooting of cutting is the main aim. It is used mostly in combination with organic materials like Farm Yard Manure or leaf mould. It provides aeration and proper drainage.

Coir Pith

Coir pith is obtained as waste from coir industry. It is composted and brought to use as soil supplement and conditioner. It is composted by the fungus

Pleurotus sojarcaju. Though it takes longer time to get composted, it is rich in plant nutrients. The percentage of N, P_2O_5 and K_2O is 0.26-1.06, 0.06, 1.2, respectively. It has calcium and certain micronutrients also. It is preffered in potting mixture as it absorbs water to the extent of five times its weight.

Leaf mould

It is produced by slow decomposition of leaves and garden sweepings. The waste leaf material is heaped in a shady place or put in a pit, dug for this purpose and allowed to slowly decay by fungal action. It has high amount of humus and water holding capacity. The nutient content depends upon the kind of leaf.

Sphagnum moss

Commercial sphagnum moss is the dehydrated young residue (or) living portion of acid-bog plants in the genus *S. papillosum, S. capillaceum and S. palustre.* Sphagnum moss is collected from the trunks of forest trees in hills (1500 MSL) during rainy season. It is relatively sterile, light in weight and has high water holding capacity and absorbs water 10-20 times of its weight. It contains specific fungistatic substances which inhibit damping off disease of seedlings. It is commonly used as rooting medium in air layering.

Compost

Compost is defined as the biological decomposition of bulk organic wastes like farm wastes like straw and bedding of animals, leaves, grass and other garden wastes under controlled condition. It takes a long time to decompose.

Farm yard manure

It is mainly of biological decomposition of cow dung of farm animal wastes. The nutrient status of animal excreta depends upon the kind of feed and fodder given to the animal. Generally it contains 0.6% N, 0.35% phosphorus and 0.6% potassium. It needs sterilization, since it has weed seeds, nematodes, insects and pathogens.

Peat

It refers to the remains of aquatic, marsh, bog or swamp vegetation which has been preserved under water in a partially decomposed state. There are three types of peat:

- Moss peat – It is derived from *Sphagnum hypnum* or other mosses. It holds water to the tune of fifteen times of its weight.
- Reed sedge peat – It is composed of remains of grasses, sedge seeds and other swamp plants. It holds water ten times its weight.

- Peat humus – It is the advanced stage of decomposition of swamp that the original plant cannot be identified. It can originate from hypnum moss or seed sedge peat. It has a low moisture holding capacity.
- Peat should be pasteurized before mixing along with other media to get rid of weed seed, insect and pathogens.

Vermiculite

This is the micaceous mineral which expands markedly when heated. Chemically, it is a hydrated magnesium-aluminium-iron silicate. It is very light in weight and absorbs large quantity of water. Expanded vermiculite should not be pressed when wet since this will destroy its desirable porous structure; it has a relatively high cation exchange capacity and thus can hold nutrients and later release them. It can be used as a rooting medium in air layering and also in potting mixtures.

Perlite

Perlite is a grey white silicaceous material. The crude ore is crushed and screened then heated to 760°C. At this temperature, the water molecules inside the particle convert to steam and expand the particle to small sponge like kernels with less weight. It holds water 3-4 times of its weight. It increases aeration, but has no buffering action, no cation exchange capacity, and no minerals. Perlite with peat moss is very popular rooting media for cuttings and tissue culture plants.

Pumice

Pumice is mostly silicon dioxide and aluminum oxide with small amount of Fe, Ca, Mg and Na. It increases aeration and drainage in a rooting mixture. It can be used alone or mixed with peat moss.

Synthetic plastic aggregates

It is a substitute for sand or perlite in high technology cultures.

- **Expanded polystyrene flakes**– These are the by-products of polystyrene processing. Resistant to decomposition. They improve drainage and aeration. They are neutral in reaction and do not absorb water and do not decay.
- **Urea formaldehyde foam** – This is prepared by mixing air with a liquid resin and allowing it to cool. It is a sponge like particle that have high water holding capacity and 30% N content, which is slowly released over several years. The materials should be used only after the loss of formaldehyde odour.

Others

Hypnaceous moss

This type of peat is partially decomposed remains of hypnum, polytrichum and other mossess of Hypnaceae family. It is suitable media for containers.

Humus or Muck

These peat materials are the decomposed debris of finely divided plant materials of unknown origin. It contains sufficient quantities of silt and clay particles and does not improve drainage or aeration. Humus is not recommended for use in growing media due to its rapid decomposition and particle size.

Bark

These are by-products of the pulp, paper and plywood industries. Hardwood bark is the common ingredient of a growing medium. Bark is leached to reduce the risk of toxicity of plants. Hardwood bark along with nitrogen makes a good potting media. Soft wood bark is acceptable but it lowers the pH of a media and so lime should be added. Barks are lighter in weight, sterile and have capacity to retain water as well as to drain the water.

Saw dust

The species of tree from which sawdust derived largely determine its quality and value for use as growing medium. The decomposed saw dust contains high cellulose and lignin content. Saw dust will be a good media, only when properly decomposed.

Bagasse

This is the waste by-product of the sugar industry. It is shredded or composted to increase the aeration and drainage properties of container media. In a growing media, the durability and longevity of bagasse is decreased and nitrogen levels affected due to its high sugar content and rapid microbial activity.

Rice hulls

These are by-products of the rice-milling industry. Rice hulls are lighter in weight and very effective to improve drainage.

Calcined clay

They are formed by heating montmorrillonitic clay minerals to 690°C. Calcined clays are six times as heavy as perlite. They have a relatively high cation exchange capacity and water holding capacity. The inorgnic soil amendments are used to increase the number of large pores, decrease water holding capacity and improve drainage and aeration.

Potting mixture: The potting mixture is prepared by using various ingredients. The proportion of each ingredient may vary with different kind of plant. While mixing, the various ingredients may be placed in a pile arranged in layers, in required ratio. Preparation of potting mixture should preferably takes place at least a day prior to use. Some dry fertilizers, as lime and super phosphate may be added during mixing. An ideal potting mixture should have an open structure which will allow good drainage and hold sufficient moisture for plant growth. It should supply adequate mineral nutrients to the plants during all stages of growth. It should be free from all harmful organisms and toxic minerals and should be light in weight. It must be firm and dense to hold the seed or base of the plant. It should not shrink on drying. Traditional potting mixture for rooted cuttings and young seedlings contains one to two parts of sand, one part of loam soil, one part of leaf mould and one part of cattle manure. For general container grown nursery plants, the potting mixture will contain one part of sand, two parts of loam soil and two parts of leaf mould. To check if the mixture is correct, the ball test is done. To this mixture, a little water is sprinkled, enough to make earth ball. Hold the mixture tightly in your fist and open your palm, you will get a lump of mixture in your hand. This lump should break when we apply medium pressure. If it breaks while opening our palm, it means that the mixture has too much sand, if it breaks with difficulty that means the mixture has too much clay or soil.

In other countries, saw dust (or) shredded bark are also mixed and used.

The following are some popularly used potting mixture formulations that are used for growing particular class of plants mentioned below:

1. Fruit plants

4 parts - red earth

2 parts - sand

2 parts - compost/manure

2. Potting of rooted cuttings and young seedlings

½ part - sand

1 part - loam soil

1 part - leaf mould

3. Annuals

3 parts - compost/manure

1 part - red earth

1 ½ part - sand

4. Bulbous plants

2 parts - red earth
2 parts - sand
2 parts - leaf mould
2 parts - compost / manure
1 part - loam soil
¼ part - charcoal

5. Palms

4 parts - leaf mould
3 parts - red earth
3 parts - sand
2 parts - loam
2 parts - compost / manure

6. Ferns

4 parts - leaf mould
3 parts - sand
2 parts - loam
2 parts - compost / manure
1 part - brick pieces
1 part - lime rubbish
½ part - charcoal.

7. Orchids

4 parts - peat moss
2 parts - brick pieces
1 part - charcoal
1 part - tank silt
1 part - leaf mould

8. Seed – sowing

3 parts - leaf mould
1 ½ part - red earth
1 ½ part - sand
½ part - powdered charcoal.

Different types of media

Cocopeat mixed with biofertilizer

Vermiculite

Perlite

Calcined clay

Coir bricks

Coir pith

Potting

The best time to pot plants is the monsoon, when the plants can endure rigorous transportation, dislocation and handling. Nursery beds are irrigated before lifting plants. Plants are lifted with ball of earth so that the root system is as intact as possible. Plants should not be pulled out in the hot sun. Hence potting and repotting should be done during mornings or evenings. Neither the roots nor the soil around the roots should be allowed to dry. Pots are filled up by placing some crocks first, then a layer of sand and finally pot mixture. Crocks are the pieces of broken pots, which are placed above the drainage hole at the bottom of the pot. They are placed in such a way that the drainage hole is not clogged and drainage is facilitated. The plant with the ball of earth is placed in the centre of the pot at the rate of one plant per pot. The plant is placed at one side of the pot in the case of rootstock plant used for inarching. Pot mixture is added around the ball of earth and pressed well, leaving 2 cm head space at the top of the pot, to avoid spilling of potting mixture during irrigation. The ball of earth should not be disturbed. The stem of the plant is set at the same height as it was in the seed bed. The potted plants are immersed gently in a tub of water and kept inside the water till air bubbles cease to come out. Then the pots are removed from water and placed under shade.

Repotting

Repotting is done when the pot mixture no longer has nutrients to sustain the plant growth. Sometimes the root fills the complete pot and this situation is called 'pot bound'. In pot bound situation, roots start emerging through the drainage hole. Pot bound plants will remain in the same stage of growth and ultimately die due to exhaustion of nutrients in the medium. Repotting is done according to the growth rate of the plant species. For slow growing plants, repotting is not required frequently. Repotting is best done in monsoons. Different sized pots and pots of different materials are used for this purpose.

At the time of repotting, it is better to wet the potted plants 24 hours earlier to facilitate easy depotting. The technique to remove the plant with a ball of earth intact is to keep the right-hand palm over the soil, allowing the stem of the plant in between the first two fingers and turning the pot upside down holding the pot at the bottom with the left hand and gently knocking the rim of the pot on the edge of table or any hard surface. The intact ball of earth will come out of the pot. In case of difficulties the pot is broken to take out the plant with ball of earth. After removing the plant from the pot, the crocks and sand at the bottom are scraped. The decayed dead, dried and twisted roots are removed. The earth ball is gently trimmed. The plant is placed in the new pot at the same height as it was in the old pot. The pot is filled with pot mixture and the soil is pressed well all round the plant. The pot with plant is immersed in water. The plants are kept under shade and watered daily. After about ten days under shade, the plants are exposed to sun gradually by keeping them under partial shade. The period of exposure is slowly increased until it withstands open sun. This process is called hardening.

Containers for potting and repotting of plants

Earthen pots

Earthen pots are made of burnt porous clay in various sizes. They are the best kind of pots since porous walls help in aeration. They have drainage holes near bottom straight sides, wider at the top than at the bottom to have the bulk of media where the spreading roots are present and to facilitate easy removal of plant with ball of earth and roots intact at the time of potting and repotting. Earthen pots should be washed thoroughly before potting, to open the clogged walls and ensure good aeration. In our country pots of varying sizes namely tube pots, ¼ size, ½ size, ¾ size, full sized pots and *thali* are commonly used.

Various sizes of pots used in potting and repotting

Size of pots	Height in cm	Diameter in cm
Tube pots	20	10
¼ pots	20	12
½ pots	20	30
¾ pots	25	30
Full size pots	30	35
Seed pans	10	45

Tube pots: Used to raise the rootstocks of mango and sapota for grafting purpose and for layering in plants like guava, rose, pomegranate etc.

¼ (or) quarter size: Used for potting very small ornamental plants. Pricking of seedlings can also done.

½ (or) quarter size: They are extensively used for growing well rooted cuttings and plants propagated by various methods.

¾ (or) three by four quarter size: Used for growing all kinds of ornamental plants

***Thali*:** Bigger sized pots preferred for growing shrubs cannas, palms, dahlias, roses etc.

Wooden Seed pan and seed boxes

Seed pans are shallow earthen pots of 10 cm high and 35 cm in diameter at the top. They have 1-3 drainage holes. Seed boxes are made of wood, 40 cm wide and 60 cm, long and 10cm deep with 6-8 properly spaced holes at the bottom. Broken pieces of crocks are placed against each of the holes and coarse sand is sprinkled to form a thin layer to prevent fine soil from clogging the drainage holes. Over this, required soil mixture is added. Very delicate seeds like *Cineraria, Begonia, Gloxinia* and other small seeds are sown in these containers.

Polythene bags

Small polythene bags of different gauge with holes punched near bottom for drainage are filled with pot mixture and cuttings of many ornamental plants, seeds of tree species, papaya, curry leaf, cashew etc. are sown. Sometimes it is also used for pricking of young seedlings.

Plastic pots

Plastic pots of round squares and rectangle shapes are used to keep indoor plants. They are light in weight and reusable.

Metal containers/ornamental pots

Tin or mud containers of different sizes and shapes with designs and drainage holes are used to keep ornamental plants. Light weight and shallow designed containers are used as hanging pots.

Ornamental pots

Mud pots and plastic pots

Tube pot

Pots of different sizes
Tube pot, ¼ size, ½ size, ¾ size and full

¼ size pot

Full size pot

Protray

Nursery hardening chamber

12

Establishment of an Orchard

An orchard is the place where different kinds of fruit crops are grown in large scale. An orchard is a long-term investment and needs lot of planning and expertise. The orchard crops are generally trees, which ultimately have a large spread, so when they are planted, the distance between plants and rows is kept keeping in mind the final spread of the tree. So, before the trees attain their full growth, some suitable crops which are of short duration, like vegetables, flowers, spices and herbs are grown as intercrops.

Fruit trees are perennial, may be tropical, sub tropical and temperate types and thus require proper environmental conditions and suitable management practices for their successful growth.

Major considerations while selecting a site for an orchard

The first and foremost thing is to buy a site for establishing an orchard

- The site should be accessible by road, so that inputs can be delivered to site, produce can be taken for delivery and it is easy for workforce to reach the place.
- Preferably close to the market.
- The soil should be of good quality, though it can be amended, it increases the orchard establishment cost.
- There should be a perennial source of good quality water.
- Should be in a place where abundant labour is available.
- Should not be close to a factory with pollutant effluents.
- The plants that suit the soil and climatic conditions should be planted.

Planning an orchard

It is of great advantage to prepare a plan of the orchard in advance, be it a home or a commercial orchard. A detailed survey of the site is carried out including the levels and a good map to scale is drawn. A full knowledge of the fruits to be grown and their cultivation is also prerequisite for efficient planning.

The guiding principles in the preparation of plan are:

- The orchard should be managed most profitably
- It should present as attractive look as possible.

The following **general principles** may be borne in mind while drafting a plan and as many of them as possible should be fulfilled. It should be recognized that not all of them can be adopted in every case.

- If the entire area is not of the same type of soil, each fruit should be allocated to the soil type it prefers.
- The irrigation sources should be marked and channels indicated along gradients with a view to achieve most economical conduct of water.
- Irrigated fruits should be close to the source of irrigation to avoid long irrigation channels and consequent loss of water during conduct.
- Tall wind breaks should be planted especially on the sides from which high winds are expected.
- There should be an adequate clearance between the wind breaks and the crop.
- Roads should be planned to occupy the minimum space consistent with economy of transport of orchard requisites and produce. The space between the wind break and the first row of fruit trees may often be utilized for roads and canals etc. with advantages.
- Drains should follow the gradient of the land, should be as straight as possible and concealed from the visitors, if possible.
- When varieties with pollen preferences are planted they should have the pollenizer in an adjacent block or in alternate rows so as to ensure good crop set.
- Fruits which ripen at the same time should preferably be grouped together to facilitate easy watching and harvesting. Assign rear areas for tall trees and the front for shorter ones will besides facilitating watching, also improves the appearance of the orchard. The orchard should in general present an aesthetic appearance so as to provide marked attraction.

The spacing adopted should be the optimum. The spacing allowed is usually such that the fringes of the trees will just touch one another curting out light but should not interlock. With in reasonable limits, closer spacing gives more yields in the earlier age. But in later life, the trees tend to grow taller than broad resulting in difficulty in pruning, spraying and harvesting. They also suffer from root competition inadequate nutrition, fewer fruits which tend to be smaller with comparatively poorer in colour development. So, adoption of closer spacing to accommodate more plants per acre proves to be a false economy in the long run. The spacing given to fruit plants depends on the following factors.

- The habit of growth of the plant: The spacing being equal to the spread of the plants.
- Rainfall: In the case of rainfed crops closer spacing is given in lighter rainfall areas than in heavy rainfall areas.
- Nature of soil: Trees on stiffer soils may be given less spacing as both their top and root spread are limited in such soils.
- The root stock: Root stock influences the spread of the trees and determines the spacing to be adopted.
- Pruning and training.
- Irrigation system.

The method of layout should be fixed in advance so that the number of plants required is worked out and arranged for.

Different features of an Orchard

While developing a land into an orchard, the following essential features should be included. While planning an orchard, the critical components like roads, office, store, implement shed, pump house, fence, wind break, irrigation system, spacing and planting materials need adequate attention.

a) Roads and foot paths

These two components should occupy minimum space for the economy of transport. The metal road in the main areas are advantageous because it is easier for the movement of vehicles like tractor or lorry to transport fertilizer, pesticides, harvested produce, large pruned material and planting materials like seeds, seedlings, layers, grafts, cuttings, etc. The clearance between wind break and first row of trees is advantageous for the road. There should be a gentle slope from the centre towards the edge of the road, to avoid stagnation of water during rainy season.

b) Office and store building

These should be constructed in the centre of the orchard for easy and proper supervision of work by the manager. A centrally located store is convenient for the labourers to take implements and tools needed for their work, to take the inputs like herbicides, pesticides, fungicides, fertilizers etc. to the field. In the store room, racks should be provided to keep the herbicide pesticide and fungicide. Wooden planks are arranged on the floor to keep fertilizer bags. The garden implements and tools are arranged in the rack. Storage bins are also kept in stores for storing the seeds and produces. In the office, racks are used to keep records and registers related to orchard management. Eg. Basic record, Stock register, Permanent Dead stock register, Temporary Dead stock register, Rainfall register, Daily Menorantum Sheet (DMS), Muster Roll, Attendance register, Forecast register, Tree register, Farm Produce register, Farm Implement register, Auction sale register, Fuel register *etc*.

c) Wells and water tanks

Planning and installation of irrigation system must be done in orchard well ahead of planting the fruit plants. It should be permanent with assured supply of required quantity of water. An orchard flourishes well under irrigation particularly during the dry months. Well or tank should be located at convenient places in different parts of the orchard at the rate of one well for every 2 to 4 hectares. Water tanks are used to store water. From the well, the water is lifted and stored in the tank and used for irrigation. Wells and water tanks are connected with concrete irrigation channels or pipes. From the tank, irrigation channels are used to take water to the field.

d) Irrigation channels

Two types of channels *viz.,* concrete and mud channels are laid out in the orchard. Concrete channel reduces water loss through seepage and the maintenance is easy when compared to mud channel. Weed growth is very less or nil in concrete channel. Channel should be laid along the gradient for most economical conduct of water. For every 30 m length of channel, 7.5 cm slope should be given.

Drip irrigation and sprinklers: Separate blocks in the orchard may be laid out with drip systems in order to deliver the water to the root zone of the crop. Fertigation tanks or venturi can also be provided if the fertilizer is applied through irrigation water. Sprinklers on the other hand, are used for irrigating the nursery plants and lawn area.

e) Manure pit

Manure pit is essential to dump the orchard leaf litter and waste materials after the harvest of the produce. Ideally two pits are required, to ensure continuous supply of considerable quantity of decomposed organic manure to the orchard. These should be located in a corner of the orchard.

f) Fencing

A good fence is essential. It may be live fence or artificial ones. Live fencing is economical and cheap, in comparison to other kind of fences. The plants suitable for live fencing should be drought resistant, easy to propagate from seed, quick growing, have dense foliage, should stand severe pruning and should be thorny. Non-thorny plants can also be used as fence plants. Non-thorny plants used as fence are Tamarind, *Thevitia, Lawsonia, Casuarina* and *Gliricidia.* Thorny plants used as fence are *Agave, Prosopis juliflora, Pithachlobium dulce, Commiphora* sp. and cactus if closely planted in 3 rows would serve as a good live fence.

In artificial fencing, stones or concrete pillars are erected at regular spacing (4 or 5 feet) and are connected by barbed wire. Recently, in many horticultural estates, solar fencing is installed to protect the crops from wild and domestic animals. Though it is relatively costly, it is very effective and is long lasting without much maintenance, when compared to barbed wire fencing.

g) Wind breaks

Wind breaks are rows of tall trees planted close together around the orchard. These are essential to resist wind velocity which cause severe ill effects particularly moisture loss from the soil through evaporation and fruit drop. Since the wind breaks are very efficient in reducing the wind velocity and minimizing the damage to the fruit trees and to other crops, their presence in regions where strong wind prevailing is of paramount importance. A wind break ordinarily has its maximum effectiveness for a distance about four times its height but its effect diminishes with distance.

The most effective wind break is a double row of tall trees alternately placed. The clearance between wind break and the first row of trees is advantageous for the road. The minimum space between the wind break and the first row of the fruit trees should be equal to tree to tree planting distance in the orchard.

It is preferable to dig a 90 cm deep trench, at a distance of 3 m from the wind break trees and prune and cut all the roots of wind break trees, exposed to sun and again fill up the trenches. This has to be repeated every 3 or 4 years in order to avoid competition between wind break and fruit trees for moisture and nutrition.

Trees suitable for wind breaks should necessarily have some special characters. They are:

- It should be fast growing.
- It should be able to acclimatize to the environment.
- It should have dense canopy.
- It should not harbour pests and diseases.
- It should be frost and drought resistant.

- It should be easily propagated.
- Planting material should be easily available and cheap.
- It should have multipurpose uses like fuel wood, fodder etc.
- It should withstand periodical pruning.

The trees which are suitable for growing as wind breaks are

- *Casuarina equisetifolia*
- *Polyalthia longifolia*
- *Eucalyptus globules*
- *Grevillea robusta*
- *Pterospermum acerifolium*
- *Azadirachta indica.*

h) Plant protection structures

These are required when planting material is maintained in a nursery for orchard

Mist chamber : For propagation of plants specially for rooting of cuttings.

Shade net : For hardening of nursery plants.

Poly house : To maintain rooted plants, before they are hardened.

i) Orchard blocks

The orchard block is the place in the orchard where the fruit trees are to be planted. When laying out a block, a base line should be drawn as per Pythagoras theorem. A baseline is drawn parallel to a permanent feature like road, fence or adjacent orchard, at a distance equal to half the spacing to be given between the trees. Pegs are fixed on the baseline at regular interval. At both ends of the base lines at right angles, lines are drawn to form a hypotenuse line, based on 3,4,5 system ($ab^2+ac^2=bc^2$). After the formation of a triangle, it is easy to fix all the other pegs to mark the tree location at required spacing using ropes to connect the pegs. This is done to avoid wastage of area and to add aesthetic value to the orchard.

When planting fruit trees, each kind of fruit should be assigned a separate block in the orchard. Fruits ripening at the same time should be grouped together. Optimum spacing should be provided to accommodate maximum number of trees per unit area.

j) Plant allocation

Short fruit trees (e.g., guava, pomegranate, annona, aonla, star gooseberry etc.) should be planted in the front and tall fruit trees (avocado, mango, sapota, jackfruit etc.) at the back in the orchard, for easy watch and to improve the appearance of the orchard.

Fruits attracting birds and animals should be close to the watchman shed, so that watchman can protect them to the maximum extent possible.

Evergreen trees (e.g., papaya, oranges, lemon etc.) should be planted in the front and deciduous ones (e.g., plum, pear, peach and apple), behind the evergreen trees.

Most of the deciduous fruit trees are self-unfruitful that means though they are bisexual, their own pollens are unable to fertilize and form fruit, consequently remain barren. To get optimum fruits there is requirement of a pollen donor which is called a pollinator variety, which pollinates and set fruits. For example, in pears, varieties 'Beurre Hardy' and 'Flemish beauty' are self-fruitful. These are used as pollinators for self-unfruitful varieties like 'Bartlett' and 'Anjou'. In cases where pollinators are required, the orchard layout should have pollinators. Ideally every third tree in every third row should be planted with a pollinator variety or every fourth tree in every fourth row should be planted with pollinator variety to enhance pollination and fruit set in these crops. It should also be planted in continuous row or placed within the row.

Pollinators placed in solid rows

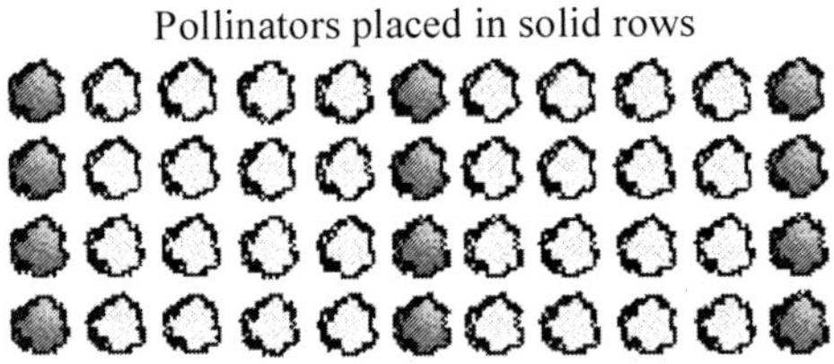

Shadded trees representpollenizers

Pollinators placed within rows

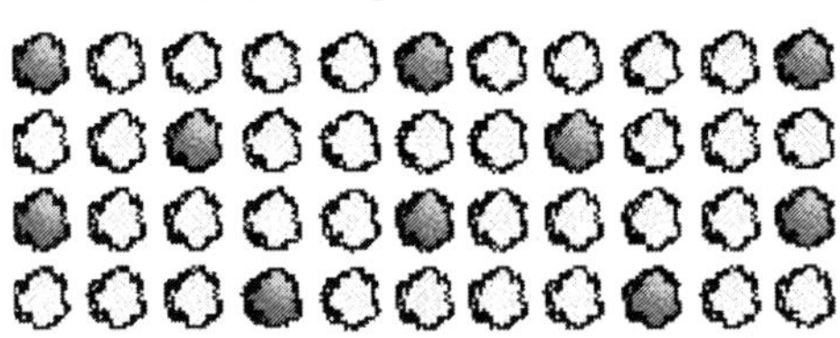

Considerations for orchard planting

While planting a new orchard, utmost attention and care is to be given to various aspects like:

- Nature of soil and subsoil.
- Selection of suitable plants and varieties after performance and market research.

- Proper pit marking at specified tree to tree and row to row distance.
- Proper pit digging and filling.
- The style of layout.
- Purchasing healthy plants of desired cultivars from reliable source.

Orchard layout

The arrangement of plants in the orchard is known as lay-out. The following points need to be considered before choosing a system of planting.

- It should accommodate maximum number of plants per unit area.
- It should allow sufficient space for the development of each tree.
- It enables equal distribution of area under each tree.
- The intercultural operations such as ploughing, spraying etc. are easily carried out.
- It makes supervision more easy and effective.

Fruit trees are perennial, may be tropical, subtropical and temperate types and thus require proper environmental conditions and suitable management practices for their successful growth. The system of layout can be grouped under two broad categories viz.

a) Vertical row planting pattern: In this method, the trees in a row are perpendicular to trees of the adjacent rows.

b) Alternate row planting pattern: Here the trees in even rows are mid-way of the trees in the odd rows.

a) Vertical row planting pattern

1. Square system

In this system, trees are planted on each corner of a square whatever may be the planting distance. The plants make a square pattern. This is easiest to lay and most commonly followed system. The central place between four trees may be used to raise short lived trees. This system has enough space and permits inter cropping and cultivation can be done in both the directions.

Advantages

1. Easy lay out.
2. Ample space between rows.
3. Inter-cultural operations can easily be done in both the directions.
4. Plants get proper space and sunlight.

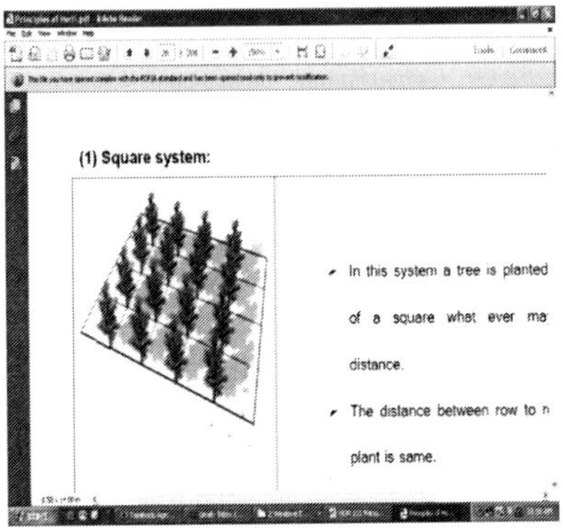

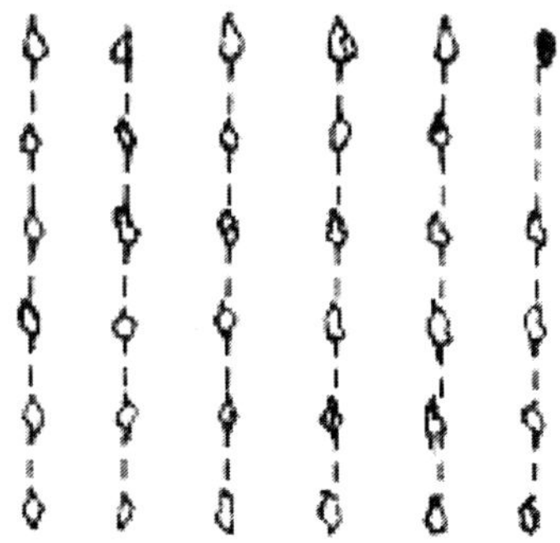

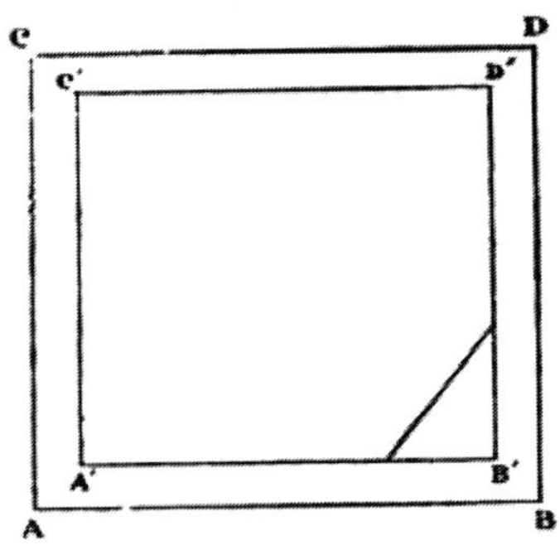

Procedure for lay out

Step no. 1: ABCD is the area where the trees are to be planted. The first step will be establishing a base line. Select the baseline parallel to the road or fence or the boundary of the orchard. This should be drawn at half a distance of the spacing that is to be followed. For example, if the spacing is 10m, the base line should be drawn at a distance of 5m from the periphery of the plot.

Step no. 2: Towards end of the base lines, leave again a gap of half the spacing from the boundary or road or fence etc. and put the peg on one end of the base line. From this peg measure one planting distance and put the second peg on the base line. Thus, continue placing pegs at each of the planting distance till the total length of the base line is covered. The distance from the last peg to the boundary should also be at half the spacing.

Step no. 3: From the first peg and the last peg on the base line, draw perpendicular lines. The perpendicular lines may be drawn by adopting any of the following methods.

Cross staff method: Cross staff comprises of a wooden block with two perpendicular slits made on its surface and fixed to an iron rod.

- Fix the iron rod in the position of the first peg. See thorough the slit parallel to the base line and see that it lies in line with the base line.
- Now see through the other slit perpendicular to the first one and fix a ranging rod or a bamboo stick at a convenient distance from the base line.
- Extend a straight line from the position of the first peg through the position of the bamboo stick. This gives a perpendicular line to the base line at the position of the first peg.

Pythagoras theorem method: Adopting a right angled triangle with the sides and hypotenuse in the proportions of 3:4:5, a perpendicular line can be drawn.

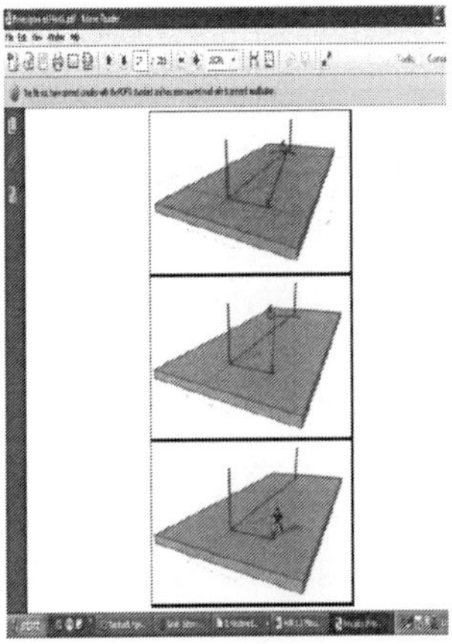

3, 4, 5 Method

- On the base line from the first peg measure a known distance in proportion of 3 and mark the point
- From the first peg, measure a distance in the proportion of 4 and draw an arc away from the base line.
- From the point previously marked on the base line now measure a distance in the proportion of 5 and draw a second arc intercepting the first one.
- Now extend a straight line from the position of the first peg through the point of intersection of the two arcs. This gives a perpendicular line to the base line from the position of the first peg.

2. Rectangular system

In this system, trees are planted on each corner of a rectangle. The row to row distance is more than the tree to tree distance. The plants make a rectangular pattern. The larger row to row distance makes the cultivation convenient in both the directions.

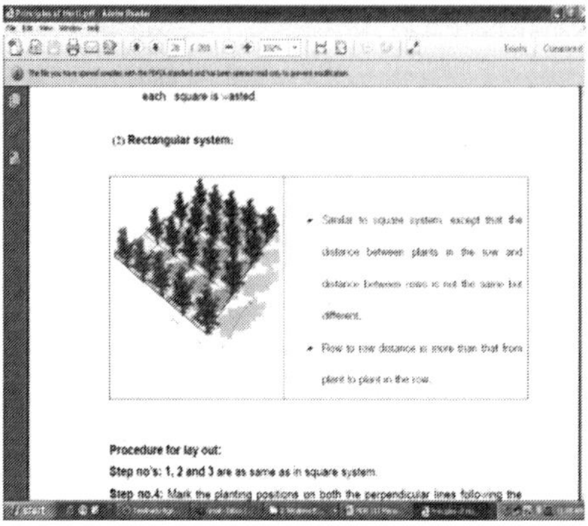

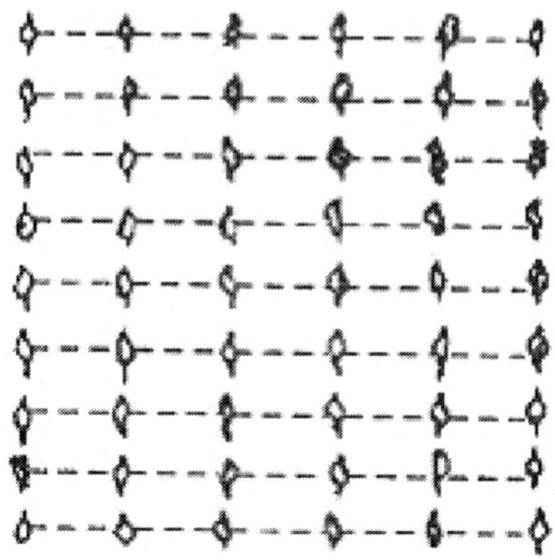

Advantages

1. More space is available between rows.
2. Inter-cultural operations can be done in both the directions.
3. Plants get proper space and sunlight.

Procedure for lay out

Step no. 1, 2 and 3 are as same as in square system.

Step no. 4: Mark the planting positions on both the perpendicular lines following the spacing to be adopted between the rows.

Step no. 5: It is same as in square system, but following the spacing to be adjusted between the rows.

b) Alternate row planting pattern

3. Triangular system

Triangular system is based on the principle of isolateral triangle. The trees are planted as in square system. The plants in the even numbered rows are midway between those in the odd rows instead of opposite to them. The distance between any two adjacent trees in a row is equal to the perpendicular distance between any two adjacent rows. However, the vertical distance, between immediate two trees in the adjacent rows, is equal to the product of 1.118 distance between two trees in a row. When compared to square system, each tree occupies more area and hence it accommodates a few trees per hectare than the square system.

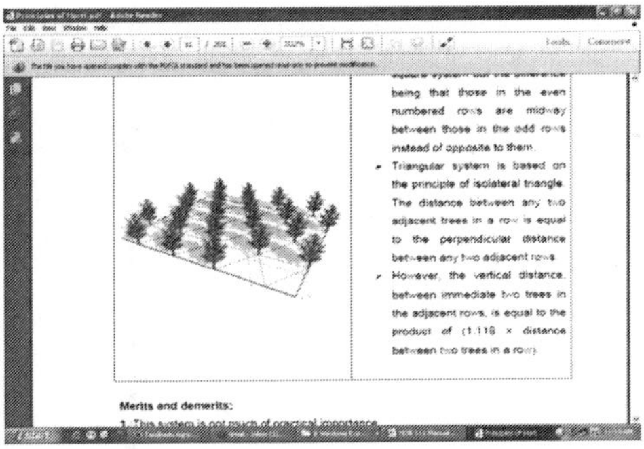

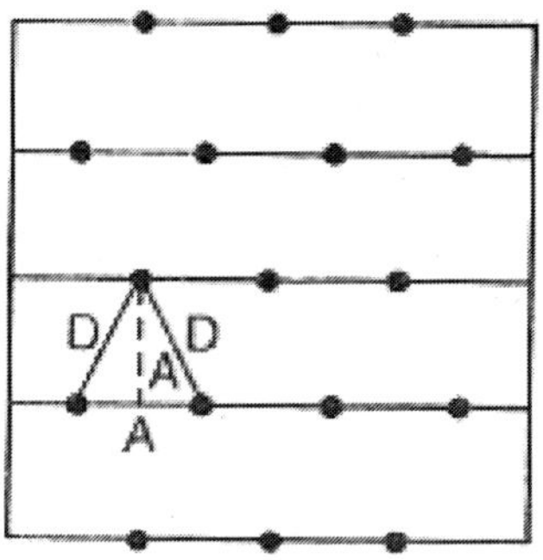

4. Hexagonal System

In this method, the trees are planted in each corners of equilateral triangles making a hexagonal pattern. Six trees form a hexagon and an extra seventh tree is planted in the centre of the hexagon. Since there are seven trees in one hexagon, this system is also called as 'septule'. This is a little difficult pattern to layout. Mathematically, perpendicular distance between any two adjacent rows is equal to the product of 0.866 x the distance between any two trees. As the perpendicular distance between any two rows is less than unity, this system accommodates 15% more trees than the square system. The disadvantages of this system are that the layout and the cultivation is also not so easy. As this system accommodates 15% more plant stand, the soil has to cater to the needs of more plants, thus fertile soil is desirable for this kind of layout.

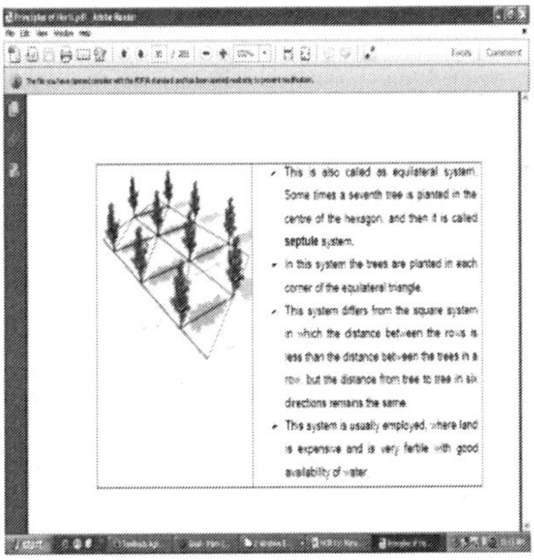

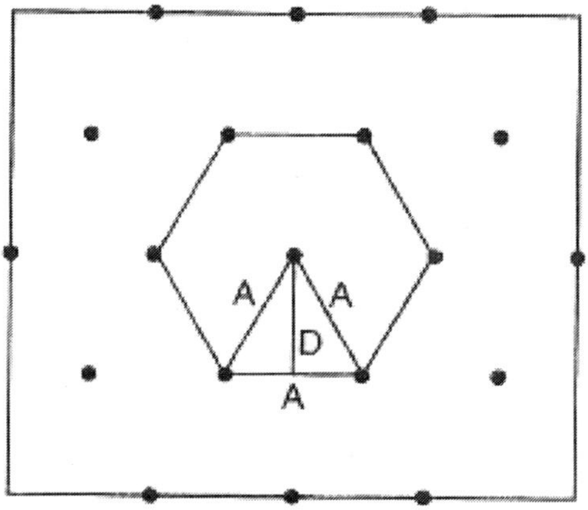

Advantages

1. Accommodates 15 % more plants than the square system.
2. Efficient use of space.

5. Diagonal or quincunx system

This is the square method but with one more plant in the centre of the square. This will accommodate almost double the number of plants, compared to the square system, but does not provide equal spacing.

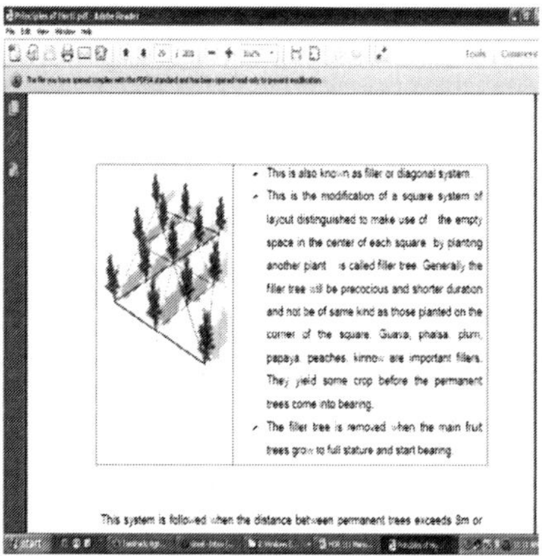

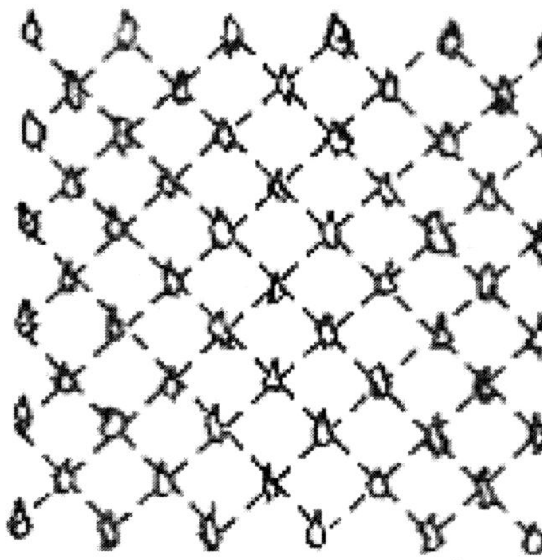

Procedure for lay out

Step no-1: Lay out the square system

Step no.-2: Draw diagonals of each square.

Step no.-3: Mark the planting position of the filler tree by fixing a peg at the point of intersection of the two diagonals in each square.

Merits and demerits

1. The main advantage of this system is that the plant population is about double than the square system.
2. The greatest disadvantage of this system is that, it is difficult to carry out intercultural operations on account of the filler tree.

Advantages

The plant population is almost double than the square system.

Contour system

- It is generally followed on the hills where the plants are planted along the contour across the slope. Trees thriving in the hill's soil and climate may be planted in this system.

- It particularly suits to land with undulated topography, where there is greater danger of erosion and irrigation of the orchard is difficult.
- The main purpose of this system is to minimize land erosion and to conserve soil moisture so as to make the slope fit for growing fruits and plantation crops.
- The contour line is so designed and graded in such a way that the flow of water in the irrigation channel becomes slow and thus finds time to penetrate in to the soil without causing erosion.
- Terrace system on the other hand refers to planting in flat strip of land formed across a sloping side of a hill, lying level along the contours.
- Terraced fields rise in steps one above the other and help to bring more area into productive use and also to prevent soil erosion.
- The width of the contour terrace varies according to the nature of the slope. If the slope becomes stiff, the width of terrace is narrower and vice-versa.
- The planting distance under the contour system may not be uniform.
- When the slope is < 10%, contour bunding is practiced and if the slope is >10% contour terracing is practiced.
- In this system the trees are planted along the contour line at right angles.
- Cultivation and irrigation can be practiced along the tree rows only.

Merits

- Undulated hilly region can be used economically.
- Soil erosion is minimised.

Demerits

- The trees may not be set at equidistance.
- The number of plants per unit area will generally be less than other system.
- Laying irrigation system is difficult in this system.
- The land preparation and planting is cumbersome.

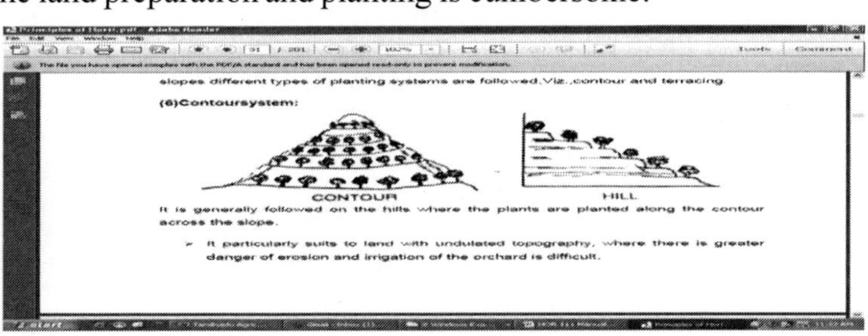

Calculation of number of plants in different systems of planting

Square system of planting $= \dfrac{\text{Area of the land}}{\text{Area occupied by a single tree}}$

Area of the land = 1 ha. (10000 sq. m^2)

Spacing between the plants and rows =10 m

Area occupied by a single tree =10 m x 10 m=100 m^2

No. of plants required per hectare= $\dfrac{10{,}000\ m^2}{100\ m^2} = 100$ plants.

Rectangular system $= \dfrac{\text{Area of the land}}{\text{Area occupied by a single tree}}$

Area of the land = 1 ha. (10000 sq. m^2)

Spacing between the plants =10 m

Spacing between the rows =12 m

Area occupied by a single tree =10 m x 12 m=120 m^2

No. of p lants required per hectare = $\dfrac{10{,}000\ m^2}{120\ m^2} = 88$ plants.

Quincunx system $= \dfrac{\text{Area of the land}}{\text{Area occupied by a single tree}} \times 2$

(Double the no. of plants of a square system)

Area of the land = 1 ha. (10000 sq. m^2)

Spacing between the plants and rows =10 m

Area occupied by a single tree =10 m x 10 m=100 m^2

10,000 m^2

No. of plants required per hectare = $\dfrac{10{,}000\ m^2}{100\ m^2} = 100 \times 2 = 200$ plants.

As the plants are planted additionally in the centre of the square, hence first, the no. of plants are calculated for square system of planting which is:

$$\frac{\text{Area of the land}}{\text{Area occupied by a single tree}} = \frac{10{,}000\ m^2}{10\ x10m} = 100.$$

Additional plants = (No. of rows length wise –1) x (No. of rows width wise-1)

In 100 x 100 Sq metre field, if planting distance is 10 x 10 m, then the number of rows length wise and width wise will be 10.

Hence, no of plants = (10-1) x (10-1) = 9 x 9 = 81.

So total no. of plants = plants planted in square system of planting + additionally planted plants in the center of the square =100+81=181.

Hexagonal system = $\frac{\text{Area of the land}}{\text{Area occupied by a single tree}}$

Method—1

In this system each tree occupies a parellogram area

Each parallelogram consists of 2 equilateral triangles.

Area of each parallelogram consists of 2 equilateral triangles.

Area of each equilateral triangle is 3/4 x a^2

Where **a** is the length of a side of a triangle ie. Spacing between the trees in a row.

So a tree occupies = **3/4** x a $^{2 \times}$ 2

Suppose if spacing is 10 m i.**e. a**

Area occupied by a single tree is **3/4** x a 2 x 2 =1.732 x 10 x 10 x 2 = 86.60 m^2

Area of the land = 10000 m^2

No. of plants per hectare = 10000 m^2

No. of plants per hectare = $\frac{10{,}000m^2}{86.60m^2}$ = 115/116.

Method-2

Suppose plant to plant distance is 10 m, then row to row distance will be calculated as under:

ABC is equilateral triangle

Hence AB=AC=BC =10 m

A perpendicular line AD is drawn on BC which divides it into two halves. It means BD=DC=5m

As per Pythagoras theory

$AC^2 = AD^2 + DC^2$ or $AD^2 = AC^2 - DC^2$

AD^2 = 102 - 52 =100-25=75.

AD = 75 = 8.66m^2

Area occupied by a single tree is = Plant to plant distance x row to row distance

=10 x 8.66=86.66 m^2

No. of plants per hectare = $\frac{10,000m^2}{86.60m^2}$ = 115/116.

Method-3

Each equilateral triangle area can be worked out by S(S-a) x (S-b) x (S-c)

Where S = half of the sum of three sides length of the triangle

a,b,c –lengths of the three sides of a triangle.

Space occupied by each plant= S(S-a) x (S-b) x (S-c) x 2

Area of the land=10000 m^2

No. of plants required = $\frac{10,000}{S\ (S-a) \times (S-b) \times (S-c) \times 2}$

High density planting (HDP)

High density aims at increasing the productivity per unit area by increasing plant population/unit area by closer spacing, without compromising on the quality of produce. Methods like using dwarfing rootstock, use of spur types of scions, mechanical operations and chemicals are used to control the size of the plant coupled with other pruning and management practices, so that production is increased. This has been successfully done in several temperate fruit crops like apple, pear, peach etc. where the desired stocks and scions are available and the plants respond to training, pruning and chemical regulation of size.

Characteristics of plants in High Density Planting

1. The trees which are to be used for high density planting system should have minimum structural branches and maximum number of fruiting branches.
2. The trees are trained with a central leader surrounded by branches which are almost horizontal.
3. The branches are pruned and maintained in such a way that each branch casts nil or least shade on other branches.

Apart from training and pruning techniques to alter the stature of plants for high density planting, certain other interventions are also considered as follows –

- Use of dwarfing rootstocks.
- Use of 'spur' types of scions.
- Use of chemicals to limit the size of the plant.

Advantages

1. Early cropping and high returns per unit area.
2. Maximum use of resources.
3. Improved fruit quality
4. Possibility of adopting mechanization.

Disadvantages

1. Competition in later years.
2. Pest and disease problems.
3. Cultural operations are difficult.
4. Limited to certain species which responds to dwarfing techniques.
5. Spur type scions, which are compatible to dwarf rootstocks can only be used, thereby the choice is limited.

In apple following types of plant densities are popular:

- Low HDP (< 250 trees/ ha)
- Moderate HDP (250-500 tree/ ha)
- High HDP (50-1250 trees/ ha)
- Ultra HDP (>1250 trees/ ha)
- Super HDP system, also called as meadow orchards

High and Ultra High-Density planting systems are now followed in tropical orchards also, e.g. mango, banana and cashewnut.

High density planting techniques developed at Tamil Nadu Agricultural University

Mango: In Neelum and Kalepad cultivars adoption of double hedge row planting system of 5m x 5m within a paired hedge and 10 m spacing between hedge rows can accommodate 222 plants per hectare as compared to just 100 plants /ha in the conventional system (10 x10 m). The yield increase was nearly 90% higher than the conventional system. This system must be integrated with proper training and pruning practices to increase the average productivity from 4.55 MT to 9.0 MT / ha.

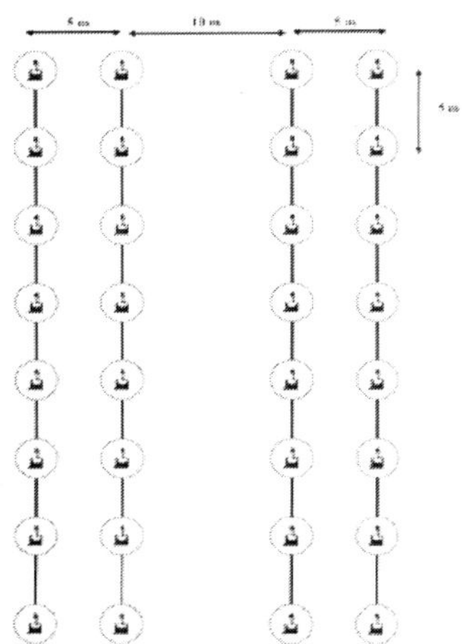

Double hedge row system in mango

Banana

Banana is grown normally at a spacing of 1.8 x 1.8 m (6' x 6') by planting one sucker per pit. HDP can be done in cv. Robusta or Grand Naine by planting three suckers/hill at a spacing of 1.8 x 3.6 m which accomodates 4600 plants/ha. In cv. Nendran, a spacing of 2 x 3 m with 3 suckers per pit (5000/ha) is recommended.

HDP in banana

Cashewnut

HDP in cashewnut

Conventionally cashew is planted at a spacing of 7 x 7 m. In high density planting system, the spacing adopted is 5 x 4 m accommodating 500 plants/ha. For sustainable high density planting, judicious training and pruning of trees from the initial years will be required. As high density planting system effectively utilizes a space available in the cashew orchards, the yield /unit area is about 2.5 times more than the normal planting density. The high density planting system when combined with drip and fertigation system, the yield potential can further be increased up to 1500 kg nuts /ha as against 640 kg/ha under rainfed condition.

Important points before taking the pits for orchard planting:

- The site ploughed twice, to fine tilth and levelled.
- Layout should be based on crops and spacing.
- Pits are dig during the summer and exposed to the sun for a period of 2-3 weeks.
- While opening the pits, the topsoil and subsoil are to be heaped separately.
- The dug-out soil is well spread on the ground, for proper exposure to sun light and drying. This is called as 'curing of soil'.

II. Method of taking pits

- For taking pits at specific planting spot, it is desirable that **planting boards** are used for accuracy of spacing.
- The planting boards are simple rectangular planks measuring 150 cm long and 10 cm wide with thickness of 3 cm.
- The planks have three notches on straight line at an equal distance.
- It is important to place the central notch on the peg hinged on the exact planting spots, as per layout.
- Subsequently, two other pegs are inserted on the remaining two notches.
- However, there is no rigidity about the direction for placing the planting board in the field.
- The pit is dug on the central notch, after removal of the planting board, keeping the other two pegs as references.
- In recent days, augers of different sizes are available which are mounted at the back of the tractor and used for taking of pits of desired dimensions precisely.

III. Pit size

- The size of the pits depends upon the nature of the soil with respect to fertility status and type of fruit plants to be grown.
- Smaller pits are prepared for quick growing fruits and smaller plants.
- Smaller pits may also be made on highly fertile soils with high water table.
- On the contrary, the larger pits are required for tall plants with large canopy.
- On less fertile soil with deeper water table, larger pits are excavated.
- Normally, the larger pits are dug one cubic meter while the smaller pits may measure 45-75 cm^3 depending on the nature of the planting and types of fruit plants to be grown.

IV. Filling of pits

- Each pit is filled with topsoil first followed by subsoil mixed with 10-50 kg well decomposed FYM + 100 g of *Azospirillum*+ 200g of AM fungi + 500g of neem cake.
- If bio-fertilizers are not available, 500 g of super phosphate may be used instead. Super phosphate is used for better rooting of saplings.
- In the top two-third portion of the pit, the proportion of the manure and soil may be kept as 1:3 ratio.

- If required use Lindane powder @100 g/pit, to control termites and ants.
- Do not mix bio-fertilizer with any chemical fertilizers/pesticides/fungicides.
- The pits should invariably be filled before the rainy season, so that there is maximum settling down of soil in the pit before the advent of heavy rainfall.
- Planting season varies from June to September i.e. planting is done during pre-monsoon period.
- The best time for planting all over India is during the monsoon season when the humidity level is high.
- In the area of heavy rainfall, the best time of planting is the end of the rainy season.
- In tracts, where the rainfall is less, the planting can be done in the early part of the monsoon season for better establishment.
- The planting should be done in the evening. If the sky is overcast, planting can be done during daytime also.

Planting Material

The planting material should be

- Healthy and vigorous.
- Devoid of pests and diseases.
- True-to-type derived from healthy mother plants .
- Certified by the concerned authorities.

Planting technique

- While planting in the pits, take care not to break the earthball of the sapling. Earthball is the ball of nursery soil, with which the sapling arrives from the nursery at the time of purchase. Breaking of earthball, disturbs the feeder roots, thus the plant becomes vulnerable and has difficulty in adjusting to new place.
- After the plant is placed in the pit, the soil should be tightened and pressed at the collar.

13

Canopy Management

Training and pruning of horticultural crops is very important practice to obtain a strong frame work of tree and sustained productivity. It is necessary to understand the various terms used to make the subject more comprehendible.

Crotch: It is the angle made by the scaffold limb to the trunk or the secondary branch to the scaffold limb.

Head: The point on the trunk from which first branch arises.

Leader: It is the main stem growing from ground level up to the tip dominating all other branches.

Low head trees: Trees in which scaffold branches arise within 60-70 cm height from the ground level.

High head trees: Trees in which the scaffold branch arise 120 cm above the trunk.

Scaffold branches: The main branches arising from the head are known as scaffold branches.

Spur: Stubby and short side stems which protrude from the main stem and bear fruit. Pruning encourages spur formation.

Trunk: The main stem of the plant.

Water shoot: It is a vigorous growing unbranched shoot arising on any branch or leader in upright position.

Water sucker: This is the growth appearing on rootstock portion.

'Training' means channelizing the growth of the tree branches in a certain way, so that the growth of the tree is acquired in a desired pattern. For example, training the tree in open centre system which means that the tree is trained to get a bowl like structure in the centre.

Objective of Training

Trees and shrubs are trained to –

- Facilitate orchard cultural operations.

- Induce balanced growth.
- Develop a strong framework.
- Provide an attractive appearance.
- Allow more light and air to the centre of the tree and to expose maximum leaf surface to the sun.
- Increase production.
- Help attain optimum colour development.
- Protect the tree trunk from sunburn injury.
- Secure a balanced distribution of fruit bearing parts on the main limbs of the plant.
- Check unproductive and excessive vegetative growth.
- Achieve optimum utilization of space.

Training is done in case of trees, shrubs and vines.

Training in fruit Crops

Training of fruit trees and shrubs

Fruit trees and shrubs are trained in a particular system depending upon their growth habit and flowering pattern. Ornamental shrubs can be trained to a particular shaped topiary like animals, birds etc. It is necessary to pay attention to the training of a plant during the first few years when the tree is young. In this period, its permanent framework is built, as decided upon by the grower.

Most deciduous and evergreen woody trees are best trained to a single stem with a low head. In case of pomegranate, custard apple and fig it would be better to train two or three stems of a plant and remove the other stems that may grow later. In the tropical climate, high-headed trees are unsuitable as the exposed trunks are subject to sunscalding in summer. Low headed trees are common all over the world. The heads or crown is kept low on the trunk allowing a distance of 60-90 cm (2-3 ft.) between the ground and the lowest branch. Such low headed trees come in to bearing much earlier, they are able to resist stormy winds more effectively, spraying and harvesting becomes easy in such trees.

Training of the branches in the main frame work of the tree is most important and the following principles are followed -

- Two to four branches, evenly spaced on the trunk are encouraged. This forms a well-balanced head.

- A stong frame work is formed, if the branches are spaced at about 15 cm distance with medium crotches (40-50).
- If two or more branches of equal size are allowed to arise from one place, they form a bad crotch and often split from their common joint.

Systems of Training of trees and shrubs

The most important systems of training are

i) Central Leader system

This system is also called as **close centre**, since the centre of the plant is closed and also as **pyramidal** system, since the plant trained looks like a pyramid. This system of training is practiced in case of certain varieties of apple and pears.

- In this system, a tree is trained to form a trunk, which extends from the surface of the soil to the top of the tree.
- In many kinds of trees, the central axis or the main branch naturally grows vertically upwards and smaller side branches grow from it in various directions.
- If the central leader is allowed to grow indefinitely, it will grow more rapidly and vigorously than side branches resulting in a robust close centre and tall tree. In such a tree the bearing is confined in top portion of the trees.

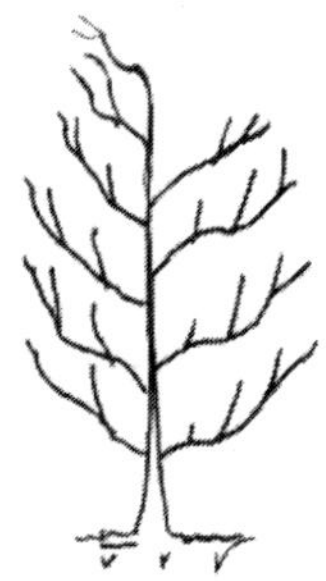

Central Leader System

Merits

- The main advantage of this system is the development of strong crotches.

Demerits

- Its main disadvantage is shading of the interior of the trees. This weakens the central leader and thus shortens the life of the tree
- Since trees are very tall, harvesting and spraying become difficult and costly.
- The lower branches, which remain more or less shaded, become ultimately less vigorous and less fruitful.
- Owing to the shading of the interior, the bearing surface moves to the periphery (outer shoots) of the tree. Thus fruiting surface is reduced and eventually yields are reduced.

- The very high shape of the plants makes them prone to wind damage.
- This method of training is not suitable for high altitude and hot arid places where wind velocity is high.

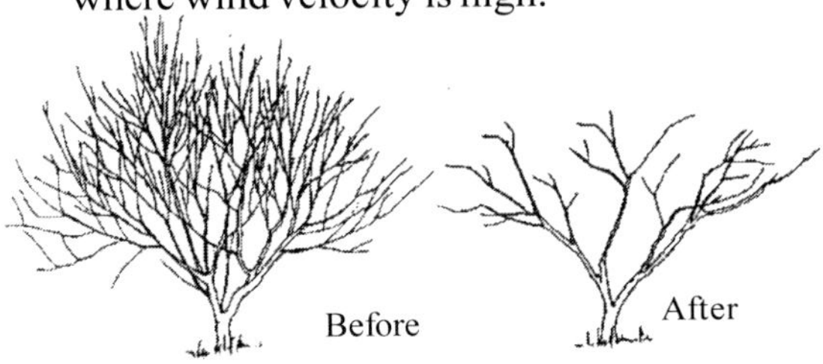

ii) Open Centre system

- In this system of training, the main stem of the plant is allowed to grow only up to a certain height and within a year of planting the plant is beheaded. Beheading of the apical portion induces lateral vegetative growth.
- This results in low head with the bulk of the crop closer to the ground.
- Since the main axis is removed, it leaves the centre open and hence **open centre** and the tree looks like a **vase (bowl),** hence it is otherwise called as **vase system**.
- This system of training is practiced in plums and peaches.

Merits

- It allows more light to reach all parts of the tree which is helpful (a) for better colour development of the fruit (b) fruiting area is spread all over the area of the trees.
- Trees become low headed thus pruning, spraying, harvesting and other cultural operations become easier.

Demerits

- The branches form weak and narrow crotches, which may frequently break under severe stress and strain such as bearing of heavy crop and strong winds.
- Sun scalding is also possible.
- The branches form very close to each other almost arising from the same spot.
- In this system the plants take the shape of a '**bowl or a vase'**, which provides a good base for setting of frost. So this system is not suitable for high altitude areas where occurrence of frost is common.

iii) Modified Leader system

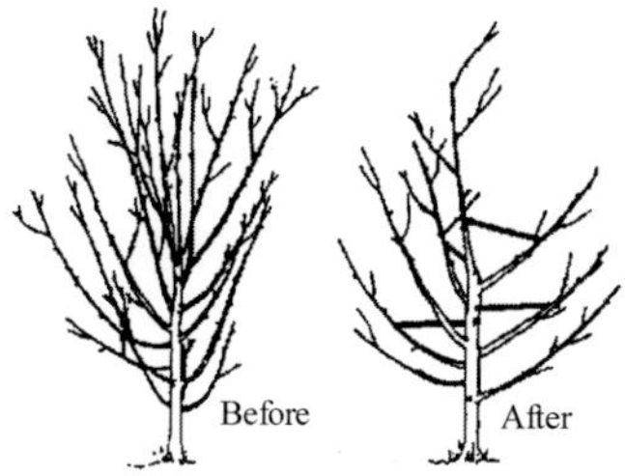

- This is intermediate system of training between central leader and open centre systems and has the advantages of both.
- This system is developed by first training the tree to the leader type allowing the leader to grow unhampered for the first four or five years.
- The main branches are allowed to arise on the main stem at reasonable intervals.
- After the required number of branches have grown, the main stem is headed back and lateral branches are allowed to grow as in the open centre system.
- This system of training is practiced in fruit plants like citrus, pear, apple and walnut etc.

Merits

- This results in a low-headed tree with well-spaced limbs, well distributed fruiting wood and low height to carryout orchard operations conveniently.
- Among the above three systems of training, the modified leader system is the best and most desirable because:
- There are narrow to medium crotches and the possibility of breakages is eliminated.
- The indeterminate growth of the central axis is also prevented keeping the tree height under control, so as to make orchard operations easy and cheap.
- No danger of sun scorching and fruit quality is good.

Systems of training for vines with support

i) Bower system

It is also called as "Pandal" or 'Arbour' or 'Pergola' system. It is generally practiced in grapes and other cucurbitaceous vegetables like snake gourd, ribbed gourd, bitter gourd etc.

- The vines are spread over a crisscross network of wires at 2.1 to 2.4m above ground, supported by concrete pillars, stone pillars or live support like *Commiphera* sp.
- The vine is allowed to grow as a single shoot till it reaches the wire network, till then it is supported by bamboo sticks tied with the jute thread.

- When the vine reaches the wires, its growing point is pinched off to facilitate the production of side shoots.
- The main advantage of this system is that it is capable of giving higher yield.
- But the main drawbacks are (i) most expensive system (ii) pruning, training and spraying operations are difficult.

ii) Espalier system

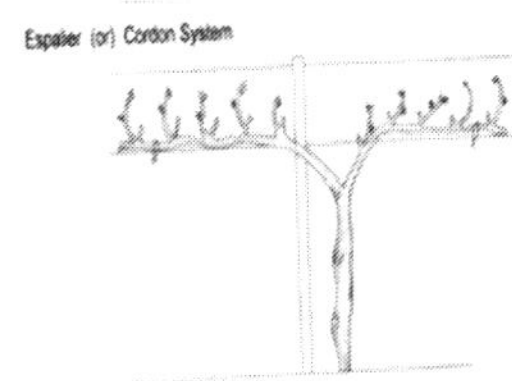

The main stem of the vine is trained to grow perpendicular on a trellis and the side branches are trained horizontally on both the sides on to wires. Plants trained in this system are called 'espaliers'. An espalier with one shoot or two shoots growing in opposite or parallel directions are called a 'cordon'.

iii) Kniffin system

In this system, two trellis of wires are supported by vertical posts. Grape when trained in this system has four canes along each wire and the bearing shoot hangs freely with no necessity of tying.

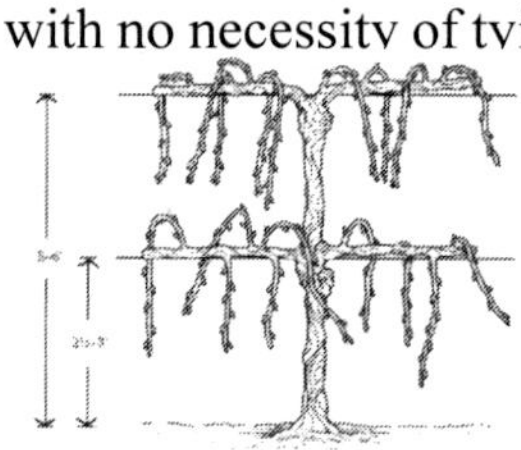

Four-arm kniffin

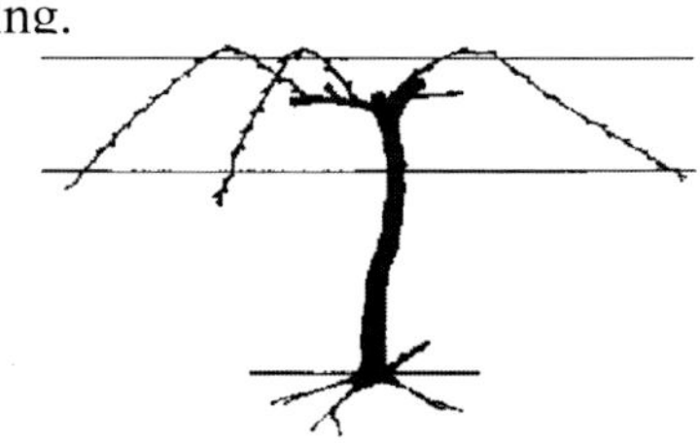

Umbrella kniffin

iv) Telephone system/over-head trellis

This system is also known as overhead trellis system. This system consists of 3 or 4 wires usually kept at 45-60 cm apart fixed to the cross-angle arms supported by vertical pillars or posts.

Telephone system

v) Tatura trellis

Vines are trained to a multi-layered wire trellis in this system. The trellis is V-shaped, supported by two long, stout poles which are embedded in soil at angles of 60º. Five wires at 60 cm intervals are fastened to these poles. This system is now being followed for pome fruits, nut fruits and grapes. The trees are grown as double leader. The leader is inclined at an angle of 60°.

Tatura trellis

vi) Y-trellis system

This system facilitates mechanization inside the grape vineyard for ploughing, bed formation, interploughing and tractor mounted spray of plant protection chemicals, foliar nutrients and plant growth regulators. It gives wide opened canopy and hence shoot positioning is better to increase grape maturity and quality. Widely adopted system in sub-tropical regions.

Y-trellis system

Systems of training vines without support

The following training systems do not require the support of any structure.

i) Head system

This system is usually followed in grapes.

- In this system, a single stem of vine is allowed to grow with the help of support of stakes.
- After attaining a height of 1.2 m, it is cut back to produce side shoots.
- Only four laterals, uniformly spaced in all directions are retained and the rest are thinned out.
- From these laterals, only two secondary shoots bearing one or two fruiting spurs are selected. As it occupies lesser area per plant, it is suitable for high density planting.

The main advantages are

- Easy and simple to train.

- Inexpensiveness to establish.
- Inter cultivation is possible.

The disadvantage is that yield obtained per plant is relatively low.

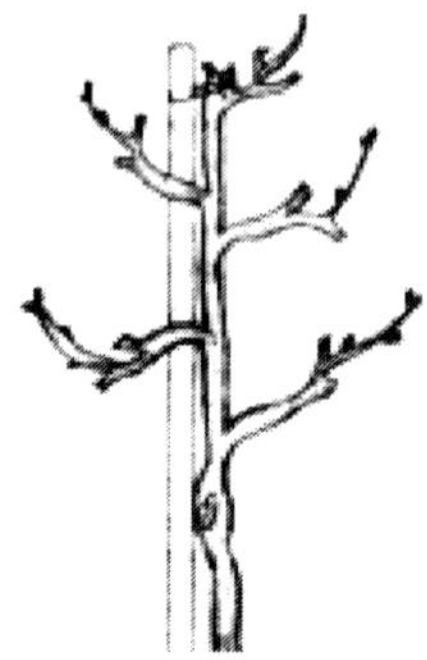

Head system

ii) Palmette

In this form of training, a central leader is encouraged, on which 3 to 4 tiers of two opposite branches are trained at an angle of 30°. Any other branches are removed.

iii) Spindle bush

This type of training consists of a central stem bearing a large number, about 15-20 semi horizontal branches radiating from it. The branches are kept spindly and no secondary growth is encouraged on primary branches.

iv) Dwarf pyramid

It is also a form of central leader, where a tree is trained with an erect central leader with 20-30 branches radiating from it on all sides at an angle of 45°-90°. The branches are trained in such a manner that the lowest branch is the longest and the topmost is the smallest, giving rise to a pyramidal shape.

v) Head and spread systems

It is another variation of modified leader system. In this system the primary branches are taken in different tiers, each tier at a spacing of 60 cm.

Head and spread system

Pruning

Pruning refers to the removal of any excess or undesirable branches, shoots, roots, or any other part of a plant so that the desired plant portion grows healthy and the plant looks according to the choice of the grower. Pruning includes removal of unwanted, surplus annual growth, dead, diseased, dried and broken branches of the plants. Pruning is an art of removing scientifically certain portions of a plant with a view to produce more and superior quality of fruit. Pruning of any kind according to its severity, changes the nutritive conditions within the tree and consequently, limits or encourages fruit bud formation.

Reasons for pruning

- Pruning helps to remove unproductive growth. It channelizes the growth of a plant in a manner in which production is more. The shade within the plant is reduced and all the branches of the plant are exposed to sun, ensuring more photosynthesis and fruit production. Pruning becomes important because if the unproductive growth is not removed, it continues to use the plant nutrients and hinders the plant performance which is reflected in low yield. If such growth is recognized and eliminated in advance, it will help in conservation of food in the tree for better production.
- Pruning removes the infected and infested portions of the plants, thereby checking the infection and infestation.
- In some fruit crops like ber, grape, guava etc., fruits are borne on current flush of growth and pruning encourages new growth, thereby producing more fruits.

Objectives of pruning

- To remove the surplus branches and direct the sap flow to the remaining branches.
- To develop a strong frame work which can carry the load of a good crop and can with stand strong winds.
- To train the plants to a definite shape e.g., Fence, Hedge, Topiary etc.
- To thin out branches so as to admit more light into the interior of the tree so that the inner wood also becomes fruitful.
- To limit the size of the tree to a convenient height so that spraying and picking of fruits is done easily and economically.
- To regulate the spacing and distribution / direction of branches.
- To distribute the fruiting wood in all directions and to maintain a balance between vegetative and reproductive phases.
- To improve the growth of the spur shoots and production of more flower buds.
- To check the further spread of the diseases.
- To maintain the vigour of the plant by removing the water shoots and other unwanted and unproductive growth.

Response of plants to pruning

The response of plants to pruning should be well understood for successfully achieving the objective of pruning. The following are some of the important ways in which the plants show response to pruning.

- **Activation of buds**

  When a branch is cut or pruned, the buds on the branch below the cut are invigorated (activated). The bud close to the cut is most vigorous and this vigour decreases in the buds as the distance increases from the cut. This is due to the elimination of the apical dominance of the terminal bud from which the auxin flows down and inhibits the growth of the lateral buds.

  This response is exploited to determine the direction of the existing branches and correcting a crotch. If the crotch is a narrow (the angle between the branch and the stem on which it arises is called crotch), the branch is pruned to a outer bud, so that the bud will produce a branch towards the outer side usually at right angle to the branch. As it grows larger, it pulls away from the stem and eventually widens the crotch. Similarly, a wide crotch can be narrowed down by pruning the branch to an inner bud.

- **Dwarfing response**

  The immediate effect of pruning is no doubt invigoration of new branches owing to the diversion of food, but due to removal of most of the foliage, there is an overall reduction in the manufacture of food, resulting in a shock on root growth. This in turn limits the further growth of the new shoots. When the growth of the new shoot is reduced, the length is also reduced. Therefore, the net effect of pruning a tree is dwarfing, which is in proportion to the severity of pruning. The spread of the top as well as the spread of the root system are reduced. This also results in dwarfing of the plant.

- **Production of water shoots**

  Severe pruning often activates the resting or adventitious buds and also buds on old wood which may sometime be stimulated to grow. They often produce branches, which grow vertically and very vigorously with long internodes, angular stems, large succulent leaves and thorns (as in citrus). They are called water shoots or water suckers or bull canes. These highly vegetative water shoots are seldom fruitful till they are several seasons old. They are wasteful and unwanted because they draw food vigorously and grow at the expense of the fruiting wood and are better removed as soon as they appear. However, they may be profitably used in some instances to fill in the gaps occurring in trees by lopsided development or loss of branches due to other causes e.g., citrus, guava, ber, sapota, cashew etc.

- **Delay in bearing**

  When pruning is severe particularly in early years of the fruit plant, bearing is delayed. Sometimes severe pruning may also lead to poor yields, because a major portion of the foliage and fruiting wood is lost.

Training and pruning system in mango HDP

First stage

Second stage

Third stage

Final stage

Methods of pruning

Thinning out

When a shoot is removed entirely from the inception (from the point of origin) so that, no new shoot arises from that place, it is referred as **thinning out**. Thinning out is practiced for the removal of shoots arising in unwanted places, water shoots etc. It is done annually. In mango, thinning out is done in August-September.

Trimming

Cutting the growth of the twigs to a pre-determined level as in the case of fence, hedge and edge.

Heading back

When the branches grow tall and vigorous without producing flowers, the shoots are headed back. When a branch is cut almost to the base, leaving a few inches of stump, carrying few buds, it is referred as **heading Back**.

The buds left on the stump give rise to shoots which are important to the tree either being spur bearers or bearing flower buds or filling up of gaps in the tree or forming vegetative wood from which flowers may arise in the following year. The shoot from the bud nearest to the cut takes the place of the pruned shoot. In mango, heading back is done every 4 to 5 years.

Pollarding

Mere cutting back of the shoots, indiscriminately to reduce the height of the tree is **Pollarding**.

Pinching (tipping)

Removal of the tip of the shoot to stop its indeterminate growth or to encourage the growth of the lateral buds is **pinching or tipping**. This is practiced in marigold, dahlia, tea and chillies at the time of transplanting.

Disbudding (nipping or rubbing)

Nipping or rubbing of young buds, preventing a chance of their sprouting is **disbudding**. When the buds arise in wrong places they are rubbed off. Similarly sprouts (buds) on root stocks are disbudded.

De-blossoming

Removal of surplus flowers to enable the tree to produce crops regularly, year after year is called **deblossoming**. This is practiced in alternate bearers like mango, apple etc. and also in crops like guava, the monsoon flowers are de-blossomed to encourage a good winter crop since monsoon crop fruits are watery and prone to infestation with fruit pest.

Seasons of pruning

- It depends on the type of wood, type of plant species and time of flower bud formation.
- Removal of disease affected, dead, and dried wood as well as water shoots can be carried out at any time of the year.
- Pruning of healthy branches should not be done when the trees are in flowering or fruiting, since the resulting disturbance leads to loss of blossoms or fruits.
- In deciduous trees, pruning can be done before the termination of dormancy.
- In ever greens, pruning should be carried out before the start of active growth or after the harvest of the crops.
- Summer pruning of deciduous trees and also the pruning of evergreens in the active growing season delays the formation of flower buds by prolonging vegetative growth.

Pruning and manuring

The sudden invigoration of a number of buds due to pruning makes a demand on the food resource of the tree. Because the new shoots are not yet ready to manufacture their own food, the reserve food in the plant may often not be sufficient to meet this demand of new growth. Further, pruning means loss of foliage and wood. So, to compensate the loss incurred due to pruning and to meet the demand of new shoots the pruned trees should be manured heavily, other wise the new shoots wither and dry. This is particularly important when old (senile) or neglected trees are pruned for rejuvenation.

Care of pruned wounds

Pruning causes wounds and these cut ends should be protected to avoid the access of disease pathogens and insect pests. So, immediately after pruning, these cut ends and wounds should be protected by applying disinfectants like Bordeaux paste or blitox paste.

Special pruning techniques

1. Root pruning
2. Ringing
3. Notching

Root pruning

Root pruning is a recognized practice for the culture of dwarf fruit trees. A circular trench which is 45 cm deep and 45 cm away from the trunck is dug and roots are to cut-off with a sharp-knife. This is done every year. After pruning, the trench is liberally filled with manure. The tree is thus fed and watered artificially in a restricted area. As a result of this, the circular mass of fibrous roots increases very slowly from year to year and the tree puts forth short but stocky and well ripened shoots. The diameter of the circular trenches may be extended slowly over the years. Each year the stumps of previous year are pruned by 4 to 5 cm. This helps to increase the production of mass fibrous roots, reduce the tree height and induce abundant fruit production. In Deccan Virdhaba area of North India, root pruning is practiced to induce flowering of oranges in required season. However, this practice is not advocated to be done every year.

Ringing

It is one of the known practices to increase fruit bud formation in certain fruit crops. The operation consists of removal of a complete ring of bark from a branch or the trunk. Ringing interrupts the downward passage of carbohydrates through the phloem and thus causes them to accumulate in the part of the tree above the ring. Ringing is practiced in mango to force flowering in over vegetative trees which do not normally bear a satisfactory crop. This practice cannot be recommended for all fruit crops and it is found beneficial in prompting fruit set in certain vigorously growing grape varieties and they often result in large sized fruits.

Notching

Notching is a partial ringing of a branch above a dormant lateral bud. The operation consists of removing a small narrow strip of bark just above or close to a dormant bud. Notching above a bud increases the yield. Notching a bud prevents the inhibitory influence of certain compounds on the bud. It alters the supply of carbohydrates coming through phloem from above and increases the supply of water and nitrogen from below through the xylem. The practice of notching below a bud is effective in starting a floral shoot. It encourages a flow of large quantity of carbohydrates in the bud and reduces the supply of water and nitrogen from below.

Ringing

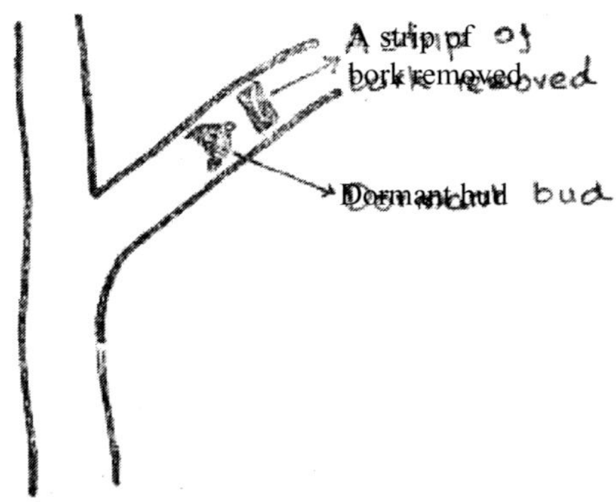

Notching

Ringing Notching

Special types of training in vegetable crops

In vegetable production some crops requires certain operations for proper development and production of good quality vegetables. Certain vegetables if left as such grow wild and do not bear abundantly unless trained or pruned in a specific form. When a plant is tied, fastened, staked or supported in a certain fashion or some of its parts are cut and removed with a view of giving the plants a frame work, the operation is called “Training”. On the other hand, “Pruning” refers to the removal of plant parts such as buds, developed shoots and roots to maintain a desirable form by controlling the direction and amount of growth.

Tomato

All indeterminate varieties are trained with wires, strings or stakes to prevent lodging and loss of fruits by coming in contact with soil. It is done by providing individual stake or by erecting 2-2.5 m long poles on either side of ridges for stretching G1 wire. Branches of plants are supported on poles or strings with a twine. Pruning is also generally followed in indeterminate varieties to improve size, shape and quality of fruits. It is the removal of unwanted shoots to enhance vigor of plants.

Desuckering

Side shoots (suckers) will develop between each compound leaf and the stem. These suckers are removed as they develop, leaving only the main stem as a growing point. Side shoots are usually not pruned until they are a few inches long, till when they are easier to distinguish from the main stem.

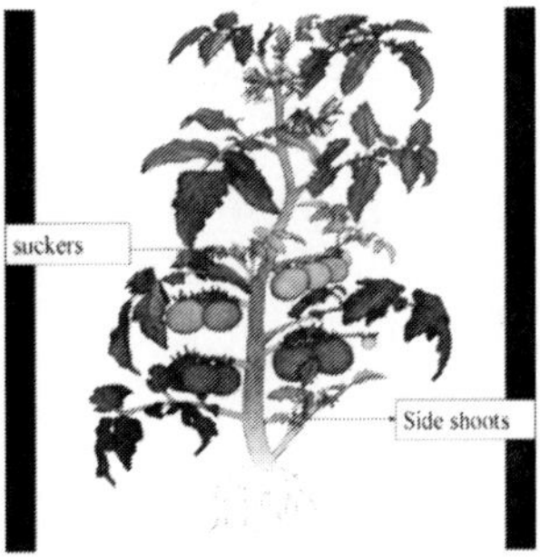

Crop support

- Nylon strings hung from horizontal wires atleast 3 m above the ground.
- A plant fully loaded with fruit weighs 7 to 14 kg.
- Each individual stem is clipped to a string with special plastic clips available in several sizes.

De-leafing

- When vines are lowered, leaves touching the ground are removed to prevent disease development.
- To increase light penetration and air circulation.
- Typically, all leaves are removed below the lowest fruit cluster.

Fruit thinning

- Small, undersized fruit at the end of a cluster (distal fruit) are to be removed for better growth and yield.

Truss hooks

Tomato, when grown under relatively low light conditions, the peduncles of the inflorescences (trusses) are found to be too weak to support the weight of fruit they bear and are liable to bend or 'kink'.

Topping

Truss hook

- Six weeks before the anticipated crop termination date, the growing point and small fruit clusters at the top of the plant are removed.
- Fruit requires 7-8 weeks from anthesis to harvest, so small fruit will not have enough time to develop to maturity.
- Topping speeds up fruit development and increases the size of already-set fruit in the lower part of the plant.
- It is helpful to leave some suckers or leaves at the top of the plant to shade the fruit and prevent sunscald.

Capsicum

Pruning

Capsicum plants are pruned to retain four stems. Pruning is done at weekly interval starting from 15 to 20 days after transplanting. Capsicum plants show dichotomous branching. The tip of the plant splits into two at 5th or 6th node 15 to 20 days after transplanting (DAT). These two branches again split into two giving rise to four branches by 25 to 30 DAT. Only these four main branches are maintained till the end of the crop. At every node, the tip splits in to two giving rise to one strong branch (the branch with thick stem and larger leaf) and one weak branch (the branch with thin stem and smaller leaf). The weak branch is removed retaining the strong branch. This operation needs to be done once in a week. From 4th month onwards, the pruning operation will be done once in 10 days. All the shoots appearing at the base of the plants are removed, as they are not productive.

All the flowers that appear till one month after transplanting are removed. Only one fruit per node is allowed to develop.

Training

The plants are trained along a plastic twine. Each branch should have a separate plastic twine to train along. The plastic twines are tied to the GI wire grid provided at 2.5 to 3.0 meter height and then to the base of the plant. The branch is tied to the plastic twine with gunny twine (3 ply). Tying plants to the plastic twine starts from 6th week after planting at 20 days interval.

Ribbed gourd

Training and pruning

In small homestead gardens, ridge gourd and sponge gourd are trailed over trellis or arbours or pandals at 1.5-2.0 m height. Commercial crops are trained on a kniffin system. The crop is trained when the seedlings are about 10-15 cm tall. The early crop can be allowed to trail on the bed itself. In Northern India, the crop is generally not trained and the vines are allowed to trail on the ground. Rainy season crop should be staked so that the fruits are prevented from direct contact with the soil.

Pumpkin

Training

For trailing pumpkin, dried twigs are spread on the ground. During the time of manuring, raking of soil should be done. Then the plants are irrigated immediately.

Cucumber

Pruning

Pruning of all the primary branches after 2 nodes gives the highest yield compared to unpruned once.

Musk melon

Training and pruning

The yield and fruit quality is improved with pruning in muskmelon. Fruits from pruned plants have higher TSS, ascorbic acid and reducing sugars and better pulp thickness than unpruned plants. The first hermaphrodite flower is borne on secondary branch arising from the eight node. The secondary branches are pinched off up to the seven node in combinations with staking is found to be more effective. Removing the secondary shoots up to the 7th node on the main stem is found optimal to improve plant growth and fruit set and to induce early flowering.

Bottle gourd

Training and pruning

In small home gardens, bottle gourd is trailed over thatched huts and walls. Various pruning methods have failed to exert favorable effects on fruit yield of bottle gourd. Bottle gourd is trained on bower systems. This increases yield over untrained vine. High productivity on bower system is due to better use of sunlight by maximum number of leaves and higher number of side branches resulting in better assimilation of carbohydrates. A spacing of 2.5 x 1.0 m is optimum.

After germination of seed, the vine is trailed on bower with the help of jute string. The axillary buds are removed weekly till the vine attains bower and finally the top of the vine is pruned 15 cm below the bower allowing two axillary buds to grow which are later trialed on the bower. Two months after sowing, the male flowers appear, following the female in the sequence of 5:5. At the end of 5th flower of female, the vines are again pruned, allowing 2-3 axillary bud to grow on the primary vines. When crop attains 2-3 months and stem of vine becomes thumb thickness, the jute string is removed and the older pale to yellow coloured leaves near the bottom of the vine are removed and destroyed.

14

Cropping Systems and Special Cultivation Practices

The types of crops and the practices used to grow them and the sequence in which they are grown altogether refers to cropping system. It comprises of all components required for the crop production and their relationships with the environment. Traditionally the cropping systems aim at maximizing the per unit area productivity, which can be done by growing improved high yielding varieties, by adequate and scientific crop management practices and by increasing cropping intensities with the help of different cropping patterns.

In today's world cropping system should produce more, at the same time the practices should be ecologically and environmentally sustainable. There should be efficient utilization of land and water. It should be designed to conserve soil, soil moisture and improve soil resilience.

Most of the vegetable crops are of short duration and fit with a number of sequences which result in greater production per unit area and time

- Choose crops that complement each other.
- Choose crops and a cropping rotation which utilise available resources efficiently.
- Keep the soil conserved.
- Choose the diverse species of crops.

Objectives

Efficient utilisation of all resources viz., land, water and solar radiation maintaining stability in production and obtaining higher net returns.

Benefits

- Maintain and enhance the soil fertility
- Enhance crop growth
- Minimize spread of disease

- Reduce risk for crop failure.
- Use resources more efficiently.

Classification of Cropping Systems

1. Mono Culture

It refers to the practice of growing a single crop, species or variety in a field, at a given time. Mono cropping of annual horticultural crops is not advisable since the crop utilizes the same kind of nutrients season after season resulting in deficiency of some nutrients, while the nutrients which are not utilized for seasons together build up to toxic levels.

Advantages

- Practiced in commercial farming and organic farming.
- Easy to plant and harvest.

Disadvantages

- Exploits one kind of resources in the area.
- Increased risk of pests and diseases.

2. Mono Cropping

Monocropping refers to the agricultural practice of growing a single crop year after year on the same land.

- Tomato, potato, banana are some common crops often grown using monocropping techniques.

Advantages of mono cropping

- Economically a very efficient system, allowing for specialization in equipment and crop production.

Disadvantages of mono cropping

- It is difficult to maintain soil cover.
- It can damage the soil ecology (including depletion or reduction in diversity of soil nutrients).
- Encourages quick buildup of pests and diseases.
- Can reduce soil fertility and damage the soil structure.

3. Multiple Cropping

It is a cropping system representing the philosophy of maximum crop production per unit area of land with in a calendar year or other relevant time span with minimum soil deterioration. Multiple cropping is a one-year cropping system in which two or more crops are grown in succession within a year. In the Garhwal region of Himalaya in India, a practice called 'baranaja' involves sowing 12 or more crops on the same plot, comprising of various types of beans, grams, and millets, and harvesting them at different times.In market garden of vegetable cultivation, the vegetable crops are cultivated on high-priced land which makes multiple cropping mandatory. In a multiple cropping system, annual and perennial plants can be organized in field together. Development of short duration hybrids has contributed to the practice of multiple cropping. Multiple cropping focuses on soil management, crop residue management, judicious use of fertilizer and other chemicals.

Advantages of multiple cropping

- Generation of high income from the farm through increased production per unit are a per year.
- Generation of more employment through intensive cropping throughout the calendar year.
- Maintenance of soil fertility and health through balanced nutrition and scientifically based crop alterations.
- Complete and economical use of natural resources, particularly nutrients and water.
- Rational use of capital and time.
- Management of pest and diseases and weeds.

Disadvantages of multiple cropping

- Provision of adequate irrigation water is mandatory. Water should not be a limiting factor in any month of the calendar year.

4. Mixed Cropping

Mixed cropping is growing of two or more crops simultaneously on the same piece of land. This type of cropping leads to an improvement in the fertility of the soil and increases in crop yield. The products and refuse from one crop plant help in the growth of the other crop plant. Mixed cropping is an insurance against crop failure in abnormal weather conditions. Planting a line of legume in-between the other crops is beneficial since the legume fixes the soil nitrogen and replenishes it. This helps the farmers to produce more and more crops without the nitrogen being depleted from the soil.

Advantages of mixed cropping

- Less fertilizer input is required.
- A single disease or pest cannot flare up and become a menace.
- In case of attack by pest or disease, the farmer does not lose the whole crop, there are alternate crops to make a living.

Disadvantages of mixed cropping

- Proper planning of the crops to be grown has to be made.
- Sowing different crops, intercultural operations and harvesting the crops individually may be cumbersome.

5. Intercropping

The practice of growing any economic crop in alley spaces of the fruit trees in the first few years or in unoccupied spaces of crops in the early periods or pre bearing stage with a view to get some extra income are known as inter crops and the practice is referred to as intercropping. The crops may or may not be sown/planted and harvested at one time.

For instance, when the main crop is cabbage, a short duration crop like radish is grown as intercrop in the available spaces and radish is harvested earlier than cabbage. When wide spacing requiring and long duration hybrids of tomato are cultivated, where the spacing is as wide as 75-90 cm, radish and lettuce are grown as intercrops, radish is harvested after 50 days and lettuce 80-85 days after sowing. Similarly, when cucumber is grown as main crop in October, cabbages or cauliflowers are grown in the interspaces successfully for initial 80-day period with a view to capitalize the advantage of early harvest in spring. Intercropping with annual crops of medicinal plants, flowers, vegetables or fruits are a common practice in fruit orchards, especially in the first few years of plant establishment.

To select and decide on the intercrops, one should know the spacing required for the plants to grow and the crop duration. Intercrops should not occupy the area where the roots of the fruit trees are concentrated. Soil fertility should be maintained or improved when intercrops are grown. Water requirements of the intercrops should not clash with those of the main crop trees. Intercrops should be selected with reference to their effect on soil moisture.

Crop	Age	Intercrop
Mango	Upto 7 years	Legumes, vegetables, papaya
Grapes	Upto 8 months	Gourds in pandal
Apple, pears	Upto 5 years	Potato, cabbage
Banana	Upto 4 months	Sunhemp, onion
Tapioca	Upto 3 months	Onion, beans, lab-lab, black gram
Turmeric	Upto 3 months	Small onion, coriander
Arecanut	Upto 10 years	Pineapple
Coconut	Upto 3 years	Banana, tapioca, vegetables

Advantages of intercropping

- More value for land and resources.

Disadvantages of intercropping

- A thorough knowledge of plant habit, duration and cultural requirement is needed to succeed.

6. Multitier Cropping

Defined as a compatible companion of crops having varying morphological frames and rooting habits grown together in such a manner that their canopies intercept solar energy at varying heights and their roots forage the soil at different zones.The main principle here is that the land, water and sunlight should be effectively used.

Tier	Crop
First (top)	Coconut / Arecanut
Second	Pepper vines trained on coconut or arecanut
Third	Cocoa or cloves, planted in the centre of four coconut or arecanut
Fourth (ground)	Pineapple, ginger or dwarf coffee

Advantages of multitier cropping

- Vertical space and resources are efficiently exploited.

Disadvantages of intercropping

- Canopy balance has to be maintained to let light pass through for the growth of lower tier plants, especially those on the ground tier.

7. Succession Cropping

The system of growing two or more crops in succession on the same land within a year. This cropping system is generally followed in most market gardens of vegetables, where the aim is to keep the high-priced land occupied with the cash crop for most part of the year. For example, in a field, vegetable cowpea is grown in June and harvested in October. This is followed by planting mid-season cauliflower in October till January and again in January onion is planted which is harvested in May. Similarly, pumpkin (October to February) is followed by okra (February to May) and brinjal (May to October). In some places, potato (October to January) is followed by onion (January to May) and okra (May to October). The crops are grown in close succession, one after the other.

Advantages of succession cropping

- The land is not kept fallow and utilized to its utmost potential.

Disadvantages of succession cropping

- The land may be rendered infertile, if crop succession is not planned properly.

8. Relay Cropping

It is also a system of growing different crops on the same land within a year but in this system, the succeeding crop is sown/planted before the preceding crop is ready for harvest. That means the growing span of two crops overlap for a short period. For example, relay of pumpkin after potato crop. Potato is harvested in March and pumpkin seeds are sown in last week of January itself and after harvesting potato, pumpkin vines are retained till June. Seedlings of early variety of cauliflower are transplanted in August and harvested in November, while the pumpkin seeds are sown in October in the standing crop of cauliflower and the crop of pumpkin terminates in February-March. Similarly, brinjal crop is grown in September through February and in December, seeds of ridge gourd/bitter gourd/watermelon are sown and the vines are retained till April. Another popular relay is of bitter gourd (October to February) and okra (January-May).

Advantages of relay cropping

- Full utilization of land.

Disadvantages of relay cropping

- The land is over exploited.
- Relay crop may be damaged if the first crop is not carefully harvested.

9. Ratooning

It is a growing practice followed in crops like banana, pineapple, chrysanthemum, where the portion of the plant which is above the ground is harvested and the roots are left in the field so that new shoot apices grow and the plant produces a fresh crop in subsequent season.

Advantages of ratooning

- Crop matures earlier.
- Cost of field preparation and planting is reduced.

Disadvantages of ratooning

- The crop may be subjected to the built up disease inoculum or pest.
- Larger dose of nutrients may be required for ratooned crop.

10. Crop Rotation

Different cropping sequences like succession cropping, relay cropping and intercropping, if practiced in the farm one after the other for two, three or more years, the cropping system may be termed as crop rotation.

The economic effect of crop rotation, succession cropping and relay cropping depends on the set of crops. However, there should be definite planning in arranging the cropping sequence based on the following principles.

Repetition of the crops having common diseases and pests should be avoided. For example, solanaceous crops should be avoided in the rotation when there is problem of *Fusarium* and bacterial wilt in tomato and brinjal. Many serious diseases and pests can be controlled by scientific crop rotation. For example, club root of cabbage and other crucifers caused by the fungus *Plasmodiophora brassicae* can effectively be controlled by keeping the land free from cruciferous crop for a period of at least three years. Similarly, root-knot nematodes (*Meloidogyne* spp., *Heterodora* spp.) can be controlled by including nematode-resistant crops like marigold, corn, wheat, rice, etc., in the cropping sequence. Rice-sweet potato-cowpea rotation has been recommended for the control of sweet potato weevil.

The relation of individual crops to their predecessors must be taken into consideration. Generally, vegetable crops of different families grow well after majority of fruit, flower or medicinal crops. But growing cucurbitaceous vegetable crops after solanaceous vegetable crops is not profitable, in general. Similarly, cabbage has marked depressing effect on the yield of onion, if onion is grown after cabbage. On the other hand, a number of vegetable crops, like onion, spinach, tomato, etc., are good forerunners for the other crops when the compatibility conditions are fulfilled.

The deep-rooted crops should be followed by shallow-rooted ones. Deep-rooted vegetable crops, like pumpkin, tomato, peas and beans, carrot, etc., are able to use nutrients extracted from deep layers of the soil, and these crops should be alternated with the crops having weak and shallow root systems, like onion, leafy greens, lettuce, cucumber, etc.

Heavy feeder crops, like cabbage, cauliflower, potato, brinjal, etc., should be followed by low-feeding crops like okra, cucumber, lettuce, pumpkin, etc.

Specific nutritional requirements of various crops should be taken into consideration. Cabbage, cauliflower, tomato, leafy greens, etc., extract large amount of nitrogen from the soil, and these crops should be followed by leguminous vegetable crops like cowpea, french bean, hyacinth bean, etc. Symbiotic nitrogen fixation by the *Rhizobium* bacteria that grows in the root nodules of the leguminous vegetable crops replenishes the nitrogen removed by the previous crop and provides benefit to the following crops.

The crops, which efficiently utilize the organic manure residues, should be grown after the crop which do not utilize fully. For that matter, immediately after the addition of organic matter to the soil, cucumber, pumpkin, summer squash, cabbage, leek etc., should be grown and the root crops, which can utilize the organic residues fully, should be followed.

Leguminous vegetable crops should be included in the cropping sequence which not only upgrade the protein status of the farm produce but also enhance soil fertility. It is to be mentioned in this context that legumes increase the susceptibility of the following crops to moisture stress because legumes are large users of water and normally leave the soil dry after harvest.

Green manuring crops should be accommodated in the rotation in order to increase the organic matter status in the soil.

Crop rotation without applying fertilizers may maintain organic matter in the soil, but does not maintain high productivity levels. A combination of crop rotation and judicious fertilizer application produce the highest yield and maintains the highest soil organic matter levels. Where previous cropping has depleted nutrients, addition of adequate fertilizers can reverse the trend of soil organic matter and nutrient depletion.

Practices like growing cover crops, mulching and growing green manuring crops is adopted to replenish the soil nutrients and save the soil moisture. These practices improve soil health as a result the crops are healthy.

Cover Crops

A cover crop is a crop planted primarily to manage soil erosion, soil fertility and soil quality. Green manure crops like sunhemp, cowpea, *Daincha* etc. are more commonly used as cover crops. Legume cover cropping in grape, mango, guava and other fruit crops is a common practice. Cowpea and french beans grow well under guava and sapota tree.In some places to prevent soil erosion, cover crops are raised mainly in the beginning or advanced rainy season so that these may provide necessary protection to the soil when damage by soil erosion is the greatest. Retention of cover crops during summer is undesirable as it will compete with the main crop of fruit trees for soil moisture. During monsoon, cover crops help in drawing off excess moisture. Berseem and methi when grown as cover crops in citrus orchards, increase the nitrogen content of the soil. Cover crops enhance soil structure by improving soil aeration and infiltration, increase the organic matter in the soil, help in pest management, suppress weeds, nematode, pest and pathogens, reduce leaching of nitrogen and reduce soil erosion and run off. *Calapogonium muconoides, Centrosema pubescens, Peuraria phaseoloides* are raised in the alley spaces and are examples of permanent cover crops.

Mulching

This is one of the most important soil management practices. Crop residues like straw, cotton stalks, leaves, saw dust, pine needles, coir dust and other materials like polythene films or certain special kinds of paper are spread in the tree basins and in inter spaces between trees. Main objective of mulching is to conserve soil moisture and to control the weed growth. It keeps soil cool in day and warm at night hours. Mulching reduces surface run-off and prevents soil erosion. It adds humus to the soil.Fruits are protected and kept clean if they fall on the mulches.It allows the absorption of more rain water and helps in bringing down the irrigation frequency.

Mulching is a system of soil management in which hay, straw, or cut grass is placed on the soil between the tree rows to conserve the soil moisture by checking surface evaporation.

It is considered that cultivation and the maintenance of soil mulch are important as it prevents the evaporation of moisture. Mulching checks germination of weeds.

Sometimes mulching may assist fire outbreak. It also facilitates the hibernation of rodents and pests. Thick mulches may act as places for mice and rodents to live and multiply. They may cause damage to tree trunks and roots by eating the bark and burrowing to the soil.

Green Manuring

Green manuring crops are primarily grown to replenish the soil nitrogen; hence the green manuring crops are leguminous.

Common leguminous green manuring crops and their potential –

Common Name	Botanical name	Green Matter (q/ha)	%N	N added in soil (Kg/ha)
Sun hemp	*Crotolaria juncea*	152	0.43	84
Dhaincha	*Sesbania aculeate*	144	0.42	77.1
Ming	*Phaseolus aurens*	57	0.53	38.6
Cowpea	*Vigna sinensis*	108	0.49	56.3
Senji	*Mellilotus alba*	206	0.51	134.4
Kesari	*Lathyrus sativus*	88	0.54	61.4
Berseem	*Trifolium alexandrinum*	111	0.43	60.7

Special Cultivation Practices

Heading back: Removal of terminal portion of shoot is termed as heading back. It is done to eradicate apical dominance and encourage the growth of new shoots.

Thinning: It is a process in which the seedlings are removed to provide optimum space for the growth of desired seedlings. This is normally done in the crops where seed nursery is sown. For example, papaya, marigold, tomato etc. Thinning helps in development of healthy seedlings. Crowded patch of seedlings hold the irrigation water and the plants succumb to damping off. In dioecious varieties of papaya, immediately after sex expression, the excess male plants are thinned.

Thinning out: Selected branches in the process of pruning to train fruit trees in a particular manner are a very important horticultural practice. It helps the plant to enable to get well distributed sun light for proper flowering and fruit development.

Girlding or ringing: A circular ring of bark of 3 cm length is removed. This hastens flowering and fruiting as it encourages the accumulation of food above the ring.

Notching: It is a method where a wedge shaped piece of bark is removed above a bud. This is done to encourage vegetative growth as notching checks the influence of inhibitory hormone. In apple, notching stimulates bud production on apical portion of the branch

Nicking: It is a method where a wedge shaped piece of bark is removed below a bud. It facilitates the accumulation of food materials through downward translocation via phloem. This encourages flowering and fruiting.

Pollarding: It is the practice of removal of top growth of the plant. This encourages new flush of vegetative growth.

Shade lopping: In plantation crops like tea, shade plants are planted in between the main crop e.g., silver oak. The branches of the shade trees are lopped every year to regulate the shade.

Bending: It is the technique followed to force flowering in guava, where the dormant shoots are bent towards the ground and tied. This helps in initiating the floral buds.

Desuckering: Removal of suckers around the main plant is called desuckering. In banana, the main plant is surrounded by small plants which emerge from the main clump. These are called suckers. The suckers have to be removed so that we can maintain healthy fruit bearing population.

Ratooning: It is the practice of allowing one sucker when the flower shoot appears in the main plant in banana so that it is ready for next cropping cycle. Normally under Indian gardenland conditions, one or two ratoons are allowed in banana. In pine apple also, after harvesting the main crown, the suckers are allowed to produce a second crop.

Propping: Staking to support banana bunches are known as propping. This prevents lodging of plant due to heavy bunch weight as banana has shallow root system and the stem is a pseudostem. This also helps to prevent damage by strong winds to some extent.

Denavelling: The female phase in banana bears fingers which remains at the top. This is followed by neutral phase when ends up in male bud at the terminal. Removal of male bud in banana after the emergence of all hands in the bunch. This helps in development of bold and healthy fingers. The male buds can be used for culinary purpose giving additional income to the farmers.

Mattocking: It is the practice of retaining 1 meter height of the pseudostem of already harvested plant, so that the nutrients accumulated in the main plant can be diverted and used in the development of the ratoon.

Detrashing: Removal of old and dried leaves in banana is commonly done for various purpose. To check the spread of leaf spot diseases. To allow proper distribution of sunlight for better performance of plant. Even the healthy leaves which rub against the fingers are removed to improve the bunch appeal. The dried leaf sheath harbours pesudostem nut weevil (*Odoiporous longicollis*). Removal of dried leaf sheaths keeps the nut weevil at bay.

Bunch covering: Immediately after the opening of last hand, at 20 days after shooting, the bunches are covered with white transparent polythene sleeves provided with 4% vent holes, which are tied in the upper end while leaving the lower end open. This cover is retained till the bunch harvest. This practiced is followed to produce unblemished appealing fruit to cater to export markets. For large scale commercial production, non-woven polypropylene sleeves are now available. These covers are reusable for 3-4 crops.

Bunch trimming: It is a practice which prevents fruits scarring and occurrence of cigar end rot in banana. The withered styles and perianth are removed from the distal end of the fruit by hand. But it is the difficult exercise to practice in tall varieties.

Pinching: The growing tip of the plant is removed. This is called pinching. Pinching helps to regulate the height of the flowering plant and gives a compact look to the plant. This is done in crops like carnation, chrysanthemum etc.

Disbudding: Sometimes a bunch of flower buds appeared which when allowed blooming, gives out a bunch of thin unhealthy flowers. In such cases, after retaining one bud, all other buds are removed. This process is called disbudding. With disbudding, healthy flowers can be obtained. It is done in rose, chrysanthemum etc.

15

Growth and Productivity Improvement

The term growth refers to an irreversible increase in size and weight of the plant. It normally reflects an increase in protoplasm, which may occur through increase in size of the cell and number of cells. Hence, the process of cell elongation and cell division provides the basis for growth. Cell division and enlargement are complicated processes involving synthesis of many organic compounds such as protein, cellulose and nucleic acid, besides the physical force required, which results in cell enlargement. The growth of the whole plant is compartmentalized into areas called 'meristem' where cell division and cell elongation are localized. The meristem cells such as shoot apex, root apex, cambium and the intercalary zones undergo repeated mitosis resulting in replication of DNA and nuclear material and finally the cell mass increases. Adjacent to this area of meristematic activity is the region of cell elongation, which phases in to the region of differentiation and maturation.

The increase in protoplasm (occurring in cell division and elongation) is brought about through a series of events in which water, carbon – di-oxide and inorganic salts are transformed into living materials. These events include (a) photosynthesis (production of carbohydrates) (b) nutrient absorption and translocation (uptake of water, nitrogen and other mineral elements) and (3) anabolism (synthesis of various materials produced in plant from the carbon fragments produced by photosynthesis and from the inorganic nutrients and water absorbed from the soil). The energy required for all these processes is provided by a catabolic process viz., respiration (degradation of sugars and fats to release energy). The net balance of anabolism and catabolism results in growth of the plant.

Development

In a plant, the development cycle begins generally with germination of seed followed by juvenility, maturity, flowering and fruiting phases. The cycle of plant growth is completed with fruiting phase and in annuals, it enters into final phase of plant growth viz., senescence and death. In perennials, the plant is ready to recycle after a period of quiescence after fruiting. The growth and development of plants consist of two distinct phases viz., 1. **Vegetative** and 2. **Reproductive phase.** The vegetative phase essentially consists of germination and subsequent development of plants and reproductive phase consist of the formation and development of flower buds, flowers, fruits and seeds or the enlargement and maturation of storage organs-fleshy stems and fleshy roots.

Vegetative phase

Germination includes all the sequential steps from the time the seed imbibes water till the seedling is self-sustaining. During germination, the complex reserve substances are converted enzymatically to simple soluble substances, which are translocated to the embryonic plant. From the stage of germination until the first flower primordium is initiated, the plant is said to be in a vegetative phase of growth. During this phase, if a plant cannot produce flowers, regardless of the environmental conditions imposed on it, then it is said to be in '**Juvenile phase**'.

The Juvenile phase is characterized by

a) Relatively rapid rate of growth

b) Distinct morphological and physiological features in some plants

e.g., In English lvy (*Hederahelix*), the leaves are 'lobed' in the juvenile phase while the leaves are 'entire' at the maturity stage. Similarly, the juvenile leaves of eucalyptus are large and broad at the base and sessile, while matured leaves are elongated and have distinct petioles.

The growth of the bud is not a continuous process but is often associated with periods of arrested development. It is of two types, one is brought about by unfavorable environmental conditions and the other one is bud dormancy, in which the growth is temporarily stopped, eventhough all the external conditions required for normal growth are conducive. The later one is more associated with internal factors, often due to the formation of growth inhibiting substances. It is broken naturally with cold temperature endured during the period of winter months.

Maturity

When a plant becomes potentially capable of reproduction, it is said to be 'mature', and this condition is often indicated by the development of flowers. Just preceding this stage, the transformation of the vegetative stem primordial into floral primordial is initiated. At this stage, distinct biochemical changes take place which alter the pattern of differentiation of vegetative bud (i.e. leaf, bud and stem tissues) to the tissues that make up the reproductive organs. At this transition stage, a distinct morphological change in the vegetative bud is also evident when examined under the microscope.

At this transition stage, growth of the central portion is reduced or inhibited and the meristem becomes flattened in contrast to the conical shape characteristics of the vegetative bud. Thus, small protuberances develop in a spiral of whorl arrangements around the meristem which subsequently develop into different floral parts. This stage of transition of vegetative buds to floral primordial is called as 'fruit bud initiation and differentiation'. The time of fruit bud initiation is very important to the horticulturists to adopt certain important horticultural practices such as pruning, application of fertilizers and irrigation etc. Many perennial plants initiate the floral primordial from few to many months before flowering as tabulated below:

Crop	Time of flower Primordia initiation
Coconut	Primordia of inflorescences begin to form in the leaf axils, about 32 months before the opening of spathe.
Mango	Flower bud differentiation takes place between August to October while flowering takes place between December to February, i.e. 5-6 months before flowering.
Banana	Fruit bud initiation and differentiation takes place between 3-5 months from planting i.e. 3-6 months before shooting (flowering) depending upon the varieties.
Sweet oranges	Flower bud differentiation takes place 3-4 months prior to flowering.
Coffee	Flowering bud initiation takes place between August – September but flowering takes place in February- March after the receipt of the summer shower.

This transition stage is influenced by the following factors

1. **Carbohydrate: Nitrogen ratio (C: N ratio)**: It has been established in many plants that if C: N ratio is optimum, it favours the flower induction, on the other hand, if nitrogen level is more, it will favour only vegetative growth and reduce or delay flowering.

2. **Photoperiodic effect:** The growth response of a plant to the length of a day i.e; length of the light and dark period is called 'Photoperiodism'. Based on the response to the light, plants can be divided into three groups viz., short-day, long-day and day neutral plants. Short-day plants initiate flowers only when the day length is below 12 hours e.g. chrysanthemum, coffee and poinsettia. Long day plants initiate flowers only if day length exceeds 12 hours e.g. beetroot, radish, spinach. Day neutral plants can initiate flowers under any day length e.g; tomato, broad bean.

3. **Moisture effects:** Moisture may influence flower initiation in certain horticultural crops.

Pollination

Pollination refers to the transfer of pollen grains from anther to the stigmatic surface. It is of two types viz., **self-pollination** and **cross-pollination**. If the transfer is from the stamen to the stigma of the same flower or to the stigma of another flower on the same plant or to the stigma of a flower on any plant of the same clonal varieties, then this type of pollination is known as **self-pollination**. If it is effected without the aid of any outside agency, such as wind or insect, then this process is known as **autogamy**. If the pollen is transferred to the flower of another individual or in the case of pomological varieties to the flower of another variety, this process is known as **cross-pollination**. Pollination is very important activity in horticultural crops, the reasons being -

1. Dioecious nature (e.g; papaya, datepalm, nutmeg) or monoecious nature (cucurbits)
2. Peculiar flower structure, (e.g; brinjal, delicious apples, vanilla).
3. Dichogamy nature (onion, carrot, sapota).
4. The behaviour of bisexual flowers as functionally unisexual (e.g. avocado and allspice).
5. Self-incompatibility or self-sterility factors (e.g. apple, buck wheat).

The pollen may be transferred from the anther to the stigma in a number of ways. In most plants, pollen is shed at or after anthesis (opening of the flower) and is transferred to the stigma by insects or by mechanical means primarily the wind. In a majority of the horticultural crops, pollination is effected by insects and pollination by wind is not common. Fruit crops like sapota, jackfruit and aonla and nutcrops like walnut, chestnut and pecan nuts are reported to be pollinated by winds. They normally produce large quantity of pollen grains which are light in weight so they can be carried to distance places. In these nutcrops, the stigma is feathery to facilitate wind pollination.

Honey bees, ants and many insects aid in cross-pollination. Their activity is greatly affected by weather conditions especially low temperature and rains. Honey bee keeping is reported to increase the fruit set and yield in many horticultural crops (e.g. apple, cardamom, coconut etc.). In apple orchards, each honey bee usually forages 2-3 trees and visits 50-100 flowers per trip, thus in a day, it visits about 5000 flowers. Hence, 10-12 colonies per hectare at a distance of 150 m are ideal for temperate fruit orchards. Recently, a chemical substance (proprietary product Name: Bee-Q) is being used to attract the bees so as to increase the fruitset. It has been estimated that nearly double the number of the flowers are pollinated in cardamom when this chemical is used at the time of early and mid-flowering phases.

Artificial pollination

Self-incompatibility is common in apple and pear. This has been overcome with the **recognition** or **mentor pollen** technique. In this technique pollination is effected with a 2:1 mixture of methanol killed or irradiated compatible pollen and self-pollens which results in seed set. It is believed that the regulatory substance from the mentor pollen to the incompatible/incongruent pollen helps to overcome the crossing barriers. Another technique **"pioneer pollen"** is also reported to increase the seed set in such fruit crops. In this method, pollination twice with compatible pollens at an interval of 1-2 days is done. It appears that the first applied pollens promote the activities of the pollens in the second application; hence the first applied pollen is called **pioneer pollen**.

In emergent situations, when adverse weather conditions prevail or the orchard has inadequate pollinizers, hand pollination can be resorted to. This is very helpful in temperate fruit orchards and is being practiced in European countries and not yet in India. The followings are some of the methods by which artificial pollination is done:

- Artificially collecting the pollens, mixing it with the spores of *Lycopodium* (fern) and applying to the flowers with a soft brush.
- Placement of bouquets – In this method, branches of flowers of pollinizers are hung in the trees to be pollinated.

Fruitset

In an orchard, all the fruit trees do not bear equally or regularly, sometimes one tree fails to bear at a given time while another tree of the same kind under similar conditions produces a heavy crop. This problem may be due to failure to set the fruits, unfruitfulness and sterility. The following terminologies are useful in understanding the problem of unfruitfulness. Fruit setting refers to the initial setting of fruit at or just after the time of blossoming and to its remaining on the plant until maturity. A plant is said to be fruitful which not only blossoms and sets fruit but carries it through to maturity, without any immature fruit drop. Plants which are unable to do this are also as **unfruitful or barren**. **Fertility** on the other hand refers to the ability of the plant not only to set fruits and yield mature fruits but also to develop viable seeds. Inability of a plant to do this is known as **infertility or sterility**. Fertile plants are fruitful and all the plants need not be **fertile. Self-fruitfulness** indicates the ability of the plants to mature fruit without the aid of pollen from some other flower, and those plants are known as **self-fruitful plants**. **Self-fertility** refers to the ability of plant of mature viable seeds without the aid of pollen from some other flower. Such plants are known as **self-fertile**.

Fruit Development

Fruit development may be divided into following phases:

1. Initiation of the fruit tissues
2. Pre pollination development
3. Post pollination growth
4. Ripening, maturation and senescence

The first two phases develop concomitantly with the flowers as a result of cell division. After pollination, cell enlargement is responsible for the major portion of size increase in most of the fruits. However, in some large fruited plants like apple, pear, water melon and squash, cell division continues even after pollination and hence, both the activities viz., increase in cell number and cell size contribute for final size.

Pollination has a vital role in the growth and development of a fruit.

a) It helps in the inhibition of flower or fruit abscission i.e. it helps in the setting of fruits.

b) It provides male gametes for fertilization.

It has been established in many cases, that pollen contains an auxin or an enzyme that converts auxin precursors present in the stigma to auxin or provides a 'synergist' effect that helps in the fruitset, but it need not result in fertilization in all cases. This is due to the fact that pollen does not germinate or even when it does, the pollen tube may burst in the style. This results in the development of fruits without seed or small seed like structures or rudimentary seeds (e.g; grapes, certain varieties of banana).

During pollination, which normally results in fruitset and fertilization, the pollen in rare cases exerts a direct influence on the matured tissues of the fruit. This phenomenon is called **'metaxenia'**. This kind of pollen affects the size, quality of the fruit colour of flesh and time of ripening e.g; Date Palms and papaya. On the contrary, the direct effect of pollen on embryonic or endosperm tissue is known as **'xenia'** e.g; temperate fruit crops such as walnut, pecan nut etc.

After fertilization, the developing fruit grows rapidly as a result of the growth stimuli (presumably auxins) received from developing seed within the fruit with the exception of parthenocarpic fruit. This has been demonstrated in many species wherein fruit size increases progressively as seed number increases from one to three or more. If an apple fruit contains only one seed on one side instead of two, that side tends to be larger. Asymmetrical fruits are often seen in stone fruits, if only one ovule develops instead of two ovules present in the ovary.

Although the control of fruit growth is located in the seed, the raw materials for development of fruits are supplied by the plant. Hence, the nutrition and moisture availability of the plant directly affect the fruit size. It has been estimated that to support the growth of one apple fruit, 40 mature leaves are required and if this ratio is altered, due to abnormally higher fruit set or excessive defoliation due to some foliar diseases, the quality and size of the individual fruits is greatly reduced.

Ripening

A matured fruit becomes ripe and senesce subsequently. Maturation may refer to the attainment of full size or the desired size for commercial use, eventhough the fruit may not be ripe. Ripening, a stage immediately after maturation includes the qualitative changes which occur after the fruit has reached full size, such as changes in pigmentation, firmness and chemical constituents. 'Senescence' is closely related to ripening but consists of the catabolic phase of fruit development during which the tissues deteriorate and organ dies.

Several processes characterize ripening

1. In some fruits, a marked rise in the rate of respiration occurs during ripening. Such fruits are called as **climacteric** e.g. apple, banana, tomato, bread fruit, papaya, peach, pear, persimmon etc. Consequently, climacteric fruits, can ripen even when detached from tree after attaining maturity.This property can be utilized for long distant transportation and reducing transportation loss. In other fruits eg. cucumber, fig. grapes, grapefruit, lemon, orange, pineapple, strawberry, etc. such sudden rise does not occur and are called **non-climacteric fruits**.
2. Colour changes often accompany ripening, this is either due to the degradation of chlorophyll or production of new pigments or due to both.
3. Fruits soften as they ripen as a result of enzymatic break down of pectins which cement the cell wall together.
4. Starch contents gets decreased and converted into sugar as ripening proceeds.
5. Ethylene gas escalates prior to the onset of ripening to the threshold level which is a prerequisite for ripening. Before ripening, fruits reach a certain stage of maturity and produce ethylene. Immature fruits lack the capacity to produce ethylene, hence do not ripen.
6. Chemicals which give the fruits their characteristics flavour, accumulate as fruits ripen. Such chemicals are volatile in nature and are present in minute quantities.

Senescence

Senescence is the erosive process that accompanies aging prior to death. After ripening of the fruit, its entire organ except the seed senesces. In the case of plant as a whole, the senescence may be **partial** or **complete**. Deterioration and death of plant organs such as leaves, stem, flowers and fruits, as in perennial plants, are known as **partial senescence** while **complete senescence** is the aging and death of the entire plant as in the case of annuals and biennials, results in termination of the life cycle.

Causes of Unfruitfulness

The causes of fertility, unfruitfulness and failure to set the fruits can be grouped into two main categories, **internal factors** and **external or environmental factors.**

I. Unfruitfulness associated with internal factors

These factors may be grouped into three -

- The evolutionary tendencies
- The genetic influences and
- The physiological factors

A. Evolutionary tendencies

Production of imperfect flowers as a cause of evolution may lead to unfruitfulness.

1. **Dioecious and monoecious nature:** Monoecious forms like melons are self - fruiting but in dioecious forms like papaya pollinizers have to be provided. It is necessary to retain a few staminate trees (ideally 10%) to ensure a good fruit set. Nutmeg which is a dioecious tree also requires pollinizers for good fruit set. The ornamental pomegranate flowers fails to set fruit. This is because they are unisexual.
2. **Heterostyly:** In certain cases, the structure and form of the flowers are peculiar in nature, which prevents self - pollination and makes cross pollination a necessity. If cross pollination does not take place, the plant is likely to remain unfruitful even though perfect sex organs have been developed. One such form is heterostyly, the style and stamens are at different heights and thus self-pollination is prevented. For example, in brinjal, four types of flowers are present, depending upon the length of style viz., long, medium, pseudo short and true short style types. The last two do not produce any fruits at all. In delicious group of apples, upright position of the stamens combined with the spreading petals permit the bees to collect the nectar without touching the pistil or anther.

3. **Dichogamy:** When the receptivity of the stigma and viability of pollen occur in different periods, this condition is known as dichogamy. This results in production of fewer fruits. If stamens ripen before the pistil, it is known as protandry (e.g. *Annona*) and if pistil ripens before the stamens, it is known as protogyny (e.g. sapota). In avocado, each flower opens twice that is, it opens once and closes, before opening again. It behaves as a female flower during first opening time and during the second time, it behaves like male flower thus preventing self-pollination.
4. **Abortive flowers or aborted pistils or ovules:** In pecan nut the terminal portion of the pistillate clusters is almost defective and is shed at pollination time. The late flowers of strawberry are always abortive.
5. **Impotence of pollen grains:** Many varieties of grapes produce impotent or non-viable pollens though they appear to have perfect flower type.

B. Genetic influence

These types of causes are due to genetic factors and are inheritable.

1. **Sterility and unfruitfulness due to hybridity:** The more distantly the parents are related, the greater is the degree of self-sterility. A hybrid between pear and quince viz., 'Purina' flowers and sets fruits freely but the fruits are seedless. Seedlessness in most of the commercial varieties of banana and pineapple is due to hybridity.
2. **Incompatibility:** It refers to the incompatibility between pollen and the ovule of the same plant or the same variety. The pollens and ovules are fertile but they fail to effect conjugation. In apple, pear and plum, incompatible varieties require another pollinizer for fruit setting.

C. Physiological influences

1. Due to slow growth of the pollen tube, e.g. cole crops.
2. Premature or delayed pollination - In tobacco flowers, if matured pollens are applied to the immature pistils, they germinate and enter the ovules but the flowers soon fall.
3. Nutritional conditions within the plant - Nutritive conditions of plant just before or at or just after the time of blossoming are important to determine the percentage of fruit set. Its effect may be on pollen viability or on fertility of the pistils. In apple, pollen from older trees are always less fertile than the pollens from young vigorous trees. In plums, exhaustion or weakening of the trees in one season by over bearing, drought and poor soil conditions is associated with the production of many defective pistils in the following year.

II. Unfruitfulness associated with external Factors

1. Nutrient supply

Sterility normally occurs in certain families due to over feeding. e.g. Graminae, Cruciferae and Leguminaceae. It has been observed that Jonathan apple, which is 'self-sterile' on rich soil and becomes 'self-fertile' in poor soils.

2. Pollinizers

In temperate fruit crops like apple, adequate number of pollinizers are required to ensure good fruit set.

3. Pruning and training

Pruning tends to produce more true hermaphrodite flowers in grape variety 'Hope'. If pruning is not done, the variety tends to remain sterile and produces aborted pistils.

4. Locality

Jonathan apple which is sterile is one location is reported to be 'self-fertile' in another location.

5. Season

'Ideal' is a hybrid grape cultivar which is 'self – impotent' early in the season but turns to be self-potent later on.

6. Temperature

High temperature at flowering will dry up the stigmatic fluid and prevent pollination. Tomato varieties do not produce any fruit when grown at high temperature i.e., when day temperature goes beyond 37°C.

7. Light

The development of stamens and petals in strawberry flowers takes place only when plants are exposed to long photoperiods.

8. Pest and diseases

Hopper in mangoes, mildews in mangoes and grapes etc., affect the development of fruits leading to poor fruitset.

9. Spraying the trees when they are in bloom

Spraying reduces the crop set during blossoming. Some of the fungicides have inhibiting effect on pollen grains. Bordeaux mixture at 200 to 10,000 ppm prevents the germination of pollen grains on the stigma.

16

Plant Growth Regulators

Plant growth regulators are defined as chemicals other than nutrients and vitamins which regulate (or) modify the physiological process of the plant. A plant hormones or phytohormones are chemical substances which are naturally produced by plant in specific organs and are translocated from site of synthesis to the site of action. They bring about specific physiological response, even in very low concentration. The quantity of synthesis of these hormones differ from plant to plant, plant internal and external factors.They are readily absorbed and these chemicals move rapidly through the tissues when applied to different parts of the plant. A plant growth hormone may be termed as a plant growth regulator, but not *vice – versa,* since other than natural plant hormones, plant growth regulators are synthetically manufactured chemicals.

Plant growth regulators have a diverse chemical composition. Based on the general response the chemical mediates in the plant system, they are classified as growthpromoters and growth inhibitors/retardants. However, a particular growth regulator may be growth promoter at a lower concentration and growth inhibitor at higher concentration.

- **Plants Growth Promoters** – They promote or induce growth in the plant system by cell enlargement, cell elongation, flowering and fruiting. Auxins, cytokinins and gibberellins are growth promoters.
- **Plants Growth Inhibitors** –They inhibit the plant growth by inducing dormancy, senescence and abscission.Ethylene and Abscisic acid inhibit plant growth.
- **Plant Growth Retardants** – The main action of these chemicals is retardation of growth. The stem elongation is retarded by preventing cell division in apical meristem. Cycocel, B9, B995 (or) Alar (or) SADH (Succinic acid 2,2-dimithy hydrazide), Phosphon – D are plant growth retardants.

Plant Growth Regulators are divided as follows

Auxins

Auxins are produced in growing tip meristem of shoot and root and is translocated to the main shoot. During reproductive phase, auxins are synthesized in flowers and developing fruits. Some synthetic chemicals having auxin like activity are Indole – 3 – acetic acid (IAA), Indole butyric acid (IBA), Naphthalene acetic acid (NAA), 2,4 – di chlorophenoxy acetic acid (2,4-D), etc.

Auxin helps in cell elongation, cell enlargement and cell differentiation. It exerts apical dominance, hence prohibits development of lateral buds and lateral growth.

Gibberellins

Synthesized by all plant organs like growing shoots, leaves and flowers, but more gibberellin is synthesized by seeds during seed development.Gibberellins stimulates tissue growth in young internodes. It elongates and stretches cell size.

Cytokinins

Root apical meristem is the primary site of cytokinin synthesis. Cytokinins are transported to the shoot for action. It helps in cell division and organogenesis of shoot and root. It interacts with auxin and produce varied responses. If cytokinin: auxin ratio is more, shoots develop, if cytokinin: auxin ratio is less, roots develop and if the amount of cytokinin is equal to auxin, callus is produced. Zeatin is naturally occurring cytokinin which is found in maize. Synthetic cytokinins are Kinetin, BA (Benzyl adenine), IBA, Ethoxy ethyl adenine.

Abscissic Acid

It is synthesized in mature leaves and roots and transported to the shoot system. It is produced more during environmental stress.It is also called 'stress hormone'. It helps the plant to deal with different kinds of environmental stress. During moisture stress, ABA induces closure of stomata. It induces dormancy and also helps the seeds to germinate.

Ethylene

It is the only gaseous growth regulator. It is also known as 'ripening hormone'. Fruits naturally produce ethylene with ripening. The other physiological processes that ethylene promotes is growth, epinasty, senescence and breaking dormancy. Ethrel is ethylene producing chemical. It is used to accerate ripening and stimulate uniform red colour in apples.

Brassinosteroids (BRs)

These are plant hormones a class of polyhydroxylated steroids which closely resemble animal steroids. Found in higher amounts in young tissues of *Brassica* species.

Cell elongation and cell division

It improves the quality and quantity of horticultural produce and helps plant combat environmental stress. It is used as natural plant strengthening substance.

Other compounds like triacontanols, morphactins (Chloroflurenol, dichloroflurenol, flurenolchlorofluren) and phenolics also have growth regulating activity.

Applications of Plant Growth Regulators in horticulture

Propagation of plants

The most common use of plant regulators in horticulture is to induce rooting in stem cuttings and in air and soil layers. Certain kind of plants may not successfully root under normal condition and with the aid of plant regulators; they can be easily made to induce rooting. The most commonly employed growth regulators for rooting are auxins like IBA, IAA, IPA and NAA auxin compounds especially IBA is of great use in rooting of cuttings. They can be applied in dust, liquid (or) paste formulation. The commercial dust formulations are **Seradix A, SeradixB** and **Rootamone**.

Cutting: Concentrations ranging from100-500 ppm are used for long dip method of treatment of cuttings for 12-24 hours and high concentrations of 10,000 to 20,000 for quick dip method for a few seconds. The concentrations differ according to the type of cutting i.e. herbaceous, semi-hard wood and hard wood cuttings. Applications in the form of dust as talcum preparation or in the form of a paste in lanolin are also used.

Layering: Another usage of plant regulators in plant propagation is in aiding rooting of air layering. When the ring of bark is removed from the stem, the growth regulators like IBA or IAA in powder or in lanolin paste is applied at the distal end of the bark-removed portion to promote root formation.

Grafting and Budding: Grafting of plants is a widely used horticultural practice of multiplying the desired genotypes in mango, citrus, ornamentals like hibiscus, rose, adenium etc. When the cambium of a root stock plant comes into physical contact with the cambium of a scion both form new xylem and phloem. Consequently, these become united and grow as one plant. Since, auxins have the property of promoting cell division of cambium these are often employed. Before grafting, either stock or scion or both are dipped in auxin solution. This promotes an early union and consequently a better success of grafted plants.

Control of flowering

The plant growth regulators are used for the regulation of flowering in certain crops. In pineapple, flowering is irregular and harvesting becomes a problem hence to regulate flower production, plant regulators are used. The treatment generally consists of pouring 50 ml of the solution containing 0.25-0.50 mg of NAA in the central core of plants. Cycocel and Alar at 5000 ppm and Ethrel at 100-200 ppm also induce flowering in mango during an off year. In *Jasminum grandiflorum,* the flowering period is extended by the application of Cycocel at 500 ppm. Flowering can also be induced in certain vegetables such as radish, beet root and carrot with the application of Gibbrellic Acid (GA_3).

Fruit set

Various growth regulators like IAA, IBA, IPA, NAA, 2, 4-D, 2, 4, 5-T and GA_3 have been found to improve fruit set in many crops. Among these chemicals 2, 4-D and NAA (Planofix) have been found in general to be most effective in increasing the fruit set. The optimum concentrations for this purpose are 10-20 ppm of auxins and 10-100 ppm of GA_3 in different crops. Spraying the flower cluster thoroughly 4-6 days after full bloom with 100 ppm GA_3 increased the fruit set in grape. It has been found that in chillies spraying of Planofix @ 1ml in 4.5 litres of water at 60th and 90th day after planting is beneficial for good fruit setting. By spraying 2 ppm of 2,4-D at the time of flowering, the fruit set increases in brinjal. A dose of 5 ppm 2,4-D can also be used for seed treatment of brinjal and this can increase the fruit set in brinjal. Application of 50 ppm GA_3 or 2 ppm 2,4-D at 15 days after planting and at flowering increases fruit set and yield.

Fruit drop

Losses resulting from pre-harvest fruit drop have long been a serious problem. In apples and pears, preharvest fruit drop is effectively checked by the application of 2,4-D and 2,4,5-T. Preharvest fruit drop in citrus is controlled with 2,4-D at a concentration of 10-15 ppm, NAA at 20 ppm and 2,4,5-T in the concentration of 15-30 ppm at pea and marble stages. Application of 20 ppm 2,4-D and 10-15 ppm 2,4,5-T in mandarins curtails the fruit drop. With 2,4-D at 10 ppm and NAA at 20 ppm fruit drop in mango is effectively prevented.

Parthenocarpy

Parthenocarpic fruit set could be induced in a number of vegetables like cucurbits, bhendi, brinjal, chillies and tomato and fruits like guava, straw berry, citrus, watermelon etc. IAA,IBA,NAA, 2,4-D and GA are effective in different crops. Application of 100 ppm GA induced complete seedlessness in grape varieties viz., Anab-e-shahi, pachadraksha etc. The problem of development of seeds in banana cv. poovan Nadu is controlled by application of 25 ppm 2,4-D in the bunches when the last hand is opened.

Fruit ripening

The plant growth regulators can be applied to hasten or delay fruit ripening. Plant growth regulators like 2, 4, 5-T at 25-100 ppm concentrations has been found to hasten the ripening in some varieties of plums and peaches. In banana, ethrel treatment at 2500 ppm induces ripening in 24 hours. Application of 2, 4-D at 16 ppm delays ripening in Washington Navel oranges. Ethephon hastens ripening in grapes. In tomatoes all the fruits on a plant do not mature and ripen at the same time, so harvesting is tedious. This is a serious disadvantage for mechanical harvesting. Ethephon applied 1-2 weeks before harvest promotes degreening and ripening of tomatoes. Application of smoke is commercially employed to hasten and ripen bananas, the active ingredient responsible being ethylene. Ethephon is also used for degreening and colour development of harvested fruits.

Fruit size and quality

Increase in berry size in Anab-e-shahi, Kismis and Bhokri varieties of grapes was reported when 40 ppm GA_3 was applied at bud and flower stages. Higher concentrations resulted in the increase in the length of grape berries.

Sex expression

Plant regulators can be used to modify the sex expression in crops. In cucurbitaceous vegetables, the production of male flowers will be always more in number than the female flowers and this sex ratio can be narrowed down by the application of 100-250 ppm ethrel four times at weekly intervals commencing from 10 to 15 days after sowing. This growth regulator not only increases the number of female flowers, but also produces female flowers at earlier nodes.

Certain plant regulators are employed to induce male sterility in crop plants which possess hermapheodite flowers so that such male sterile plants can be used as a female parent in the hybridization work. This process dispenses the expensive work of emasculation of anthers. Complete male sterility in bhendi can be obtained by spraying with 0.4% of Maleic Hydrazide. A single spray one week before floral bud initiation induces male sterility for 10 days and a subsequent spray at floral initiation extends the effect to 22 days.

Storage and shelf life

Long duration storage of onion bulbs is a difficult task because the bulbs sprout while storage. To arrest the sprouting during storage, a pre harvest spray of 2500 ppm Maleic Hydrazide, 15 days before harvest should be given.

Yield

Spraying 250 ppm ethrel, five times at monthly interval, starting from third month after planting can improve the tuber yield of topioca. Ethrel spray at 250 ppm, five times at 15 days interval starting from 15 days after planting increases tuber yield in sweet potato.

Preparation of growth regulators

Solution form: To prepare an alcoholic solution of any plant growth regulator, 1g of growth regulator is dissolved in 50 ml of ethyl alcohol or methyl alcohol or methylated spirit and then diluted with an equal volume of water to make 100 ml of solution containing 10,000 ppm of growth regulator. This acts as stock solution for further dilutions with distilled or de-ionized water. This is in airtight bottles in a refrigerator, the solutions retain their activity indefinitely.

Dilute solution method (soak method)

The basal part of cutting is soaked in dilute solution of IBA (20 ppm) for 24 hr prior to planting for easy rooters and 200 ppm for shy rooters.

Concentrated solution method (Quick dip method)

The concentration of compounds varies from 1000 to 10,000 ppm and the base of the cutting is dipped for few seconds before planting. The remaining solution may be reused by storing in dark for about a week. High concentration dip for longer period leads to yellowing and dropping of leaves and the cuttings turn black and die.

Dust form: The active ingredient of IBA is dispersed in talcum powder to get required strength of the dust formulation. While treating the cuttings, a fresh cut is made at the base and the growth regulator dust is immediately applied to the cut. The moisture exuding out of fresh cut will help proper adherence of dust over the cut surface. The cuttings are gently tapped with fingers to remove the excess dust and planted for rooting. Cuttings can also be treated by tying in bundles. The polarity of cutting should be checked before tying for growth regulator treatment. To prepare a dust containing 10,000 ppm of growth regulator, 1g of the growth regulator is dissolved in 40 ml of 95% methylated spirit and this stirred into 100g of pharmaceutical talc to form a smooth paste. This should be done in a dark room away from strong light. The paste is constantly stirred while it is drying until it becomes a fine dry powder. This prepared dust remains active for six months or more if stored in a closed opaque container in a refrigerator. This stock is diluted to optimum concentration before use.

Lanolin pastes: These are particularly convenient for use in air layering but now regarded as an obsolete treatment. The growth regulator is stirred into the molten lanolin and allowed to cool. To make a paste containing 500 ppm of growth regulator, 200 g of lanolin is mixed and thoroughly stirred and 1 g of required growth regulator is mixed. This prepared paste will keep indefinitely if stored in an air tight opaque glass vessel in a refrigerator.

Preparation of stock solution

1 mg in1 lit (1000 ml) of water = 1 ppm

1000 mg in 1 lit (1000 ml) of water = 1000 ppm

1000 mg (1 gm) in 100 ml = 10,000 ppm = (1 per cent)

If higher concentration is required, either the quantity of Plant Growth Regulator can be increased or the volume of solvent is reduced.

e.g; If 2000 ppm concentration is required

2000 mg in 1000 ml = 2000 ppm

1000 mg in 500 ml = 2000 ppm

$$\frac{10^6 \times \text{wt in g}}{\text{Vol. in ml}} = \text{Conc. in ppm}$$

$$\frac{10^6 \times \text{wt in g}}{\text{Vol. in ml}} = \text{Conc. in ppm} = \frac{10^6 \times 2\text{g}}{1000} = 2000 \text{ ppm}$$

$$\frac{10^6 \times \text{wt in g}}{\text{Vol. in ml}} = \text{Conc. in ppm} = \frac{10^6 \times 1\text{g}}{500} = 2000 \text{ ppm}$$

$$\frac{10^6 \times \text{wt in g}}{\text{Vol. in ml}} = \text{Conc. in ppm} = \frac{10^6 \times 1\text{g}}{100} = 10{,}000 \text{ ppm}$$

Dilution from stock solution

Stock solution of higher concentration can be prepared and from that required quantity may be measured and diluted to get required concentration.

(Eg.) Assume stock solution of 5000 ppm

500 mg dissolved in100 ml = 5000 ppm

If lesser concentration of 500, 150, 100 ppm are to be prepared follow the given procedure.

To prepare 500 ppm

Pipette out 500 ml and add water to make up to 5000 ml

$\mathbf{V_1 \times N_1 = V_2 \times N_2}$

This dilution is automatic and accurate.

Precautions

- Growth regulators should be applied or sprayed in the evening or morning hours.
- Application by high volume hand operated sprayers is recommended.
- Growth regulators should be applied in proper concentrations to achieve desired results.
- Good water should be used to dissolve the chemicals.
- If the chemical is difficult to dissolve in water at room temperature, warm water or alcohol may be used.

17

Protected Cultivation

Greenhouses are frames or inflated structures covered with a transparent or translucent material in which crops can be grown under partially controlled or totally controlled conditions of atleast. It is large enough to permit persons to work within, to carry out cultural operations. Greenhouse cultivation as well as other modes of controlled environment cultivation has been evolved to create favourable microclimate, for the crop production all through the year or during off season, when the produce is highly priced and remunerative. Green house is a broader term of protected structures which includes different types of structures according to the need. Green houses vary in size and complexity from small home or hobby structures to large commercial units, from solar green house to sophisticated environment controlled ones.

Types of protected structures

1. **Mist chambers:** The structures made up of High Density Polyethylene (HDPE) films, intended for propagation of plants, mainly for rooting of cuttings. High level of relative humidity is maintained inside the mist chamber in order to facilitate ease of rooting especially in difficult to root plants. It is also used for initial hardening of tissue cultured planting materials. Intermittent mist systems are widely used for rooting of soft wood, semi-hardwood, hard wood and herbaceous cuttings. Mist sprays provide a film of water over the cuttings and media. Intermittent mist controls water loss from cuttings by reducing both leaf and surrounding air temperature via evaporative cooling and raising relative humidity. Increase in relative humidity prevents desiccation of cuttings and provides more favorable environmental condition for root formation. As the humid condition facilities root formation in cuttings and layers, plants are usually propagated in the monsoon. Plants, which fail to root from cuttings or develop low percentage of rooting under ordinary condition, have shown satisfactory rooting under mist.
2. **Hot beds:** This may be primitive type of protected structure originally designed in Western temperate countries in order to produce plants during chilling winter. A hot bed is a bed of soil enclosed in a glass or plastic frame. It is heated by electricity, steam, or hot water pipes. Hot beds are used for

forcing plants or for raising early seedlings. The standard size of hot frame is 0.9 x 1.8 m. If polyethylene is used as the covering, any convenient dimensions can be used. Plastic and PVC tubing with recirculation of hot water are used for providing bottom heat in hot beds. Seedlings can be raised and leafy cuttings can be rooted in hot beds early in the season. For small propagation operations, these structures are suitable for producing many thousands of nursery plants. Hot bed basically is a heated cold frame.

3. **Cold frames:** A cold frame is a bottomless box with a removable top. No artificial heat is used inside a cold frame. The soil inside the box is heated during the day and gives off its heat at night to keep the plants warm. The installation of mist line or irrigation provision in cold frame is essential to maintain humid conditions. These structures are used in conditioning or hardening of rooted cuttings or young seedlings before transplanting to field, nursery row or container planting. The temperature difference between inside and outside of the **frame** is generally not more than 5 to 10 degrees. A mat or blanket may be placed over the **frame** on **cold** nights to conserve heat, but this increases temperature by only a few degrees.

4. **Glass house:** Made up of glass, which retains the same temperature for long time. It provides higher air infiltration rate which leads to lower interior humidity and better disease prevention.

5. **Poly house:** A polyhouse is a framed structure covered with polythene sheets. The size of the polyhouse determines the choice of the crops to be raised. This is used either for nursery preparation or crop cultivation. Mainly intended for cultivation of temperate or rain sensitive crops like tomato, chillies, capsicum and cucumber in the off-season or for year round production.

6. **Lath house:** This type of special structures is suitable for hot climates. The structures are covered with narrow wood slats or brushwood in order to reduce the scorching sunlight and also provide protection to the growing plants against strong wind and slight frost. Lath houses are originally popular in the Southern states of America, Mexico and many other hot climates of the world.

7. **Shade net house:** Shade net house is a framed structure made of materials such as GI pipes, angle iron, wood or bamboo. It is covered with plastics net (Nets are made of 100% polyethylene thread with specialized UV treatment) having different shade percentages. It provides partially controlled atmosphere and environment by reducing light intensity and effective heat during day time to crops grown under it. Shade nets are available in different shade percentages or shade factor i.e 15%, 35%, 40%, 50% 75% and 90%. Example, 35% shade net means, the net will cut 35% of light intensity and

would allow only 65% of light intensity to pass through the net. Each plant has its individual requirements for sunlight and shade under which it flourishes at its best. To create optimum climatic conditions, selection of the correct percentage of shade factor plays an important role to enhance plant's productivity to its optimum.These structures are generally used in sub-tropical and tropical climates to reduce light level and provide cooling. Shade houses are often used in the production of nursery stock. Shade houses are used to reduce adverse effect of scorching sun and rain for growing nurseries during summer season. They provide protection against wind, hail storm and also reduce water requirement.

8. **Insect proof net house:** The main aim of this type of structures is only to provide protection against insects and sucking pests. The framed structures covered with insect proof white nets in order to prevent the crops being affected by sucking pests thereby reducing the incidence of viral diseases. When compared to shade net, these type of net houses are ideal to grow plants as it allows nearly 90% of the sunlight, so that plants can be grown inside as like growing in the open fields. Other advantage is growing crops for seed production under isolated insect proof conditions.

Fruit crops like papaya can be grown inside these net houses and the plants found free of Papaya Ring Spot Virus.

Advantages of greenhouse

- Crops can be grow throughout the year viz., protects the crop from rain, hail stones, storms, wind and frost in winter.
- Crops can be grown in off-season in protected structure.
- Higher production per unit area is obtained by utilization of horizontal and vertical space e.g., vertical tier system of crop production.
- High value horticultural crops for export purpose can be grown. e.g., tomato, capsicum, cucumber, rose, carnation etc.
- Consumption of irrigation water and fertilizers is minimized.
- Free from infestation of pests and diseases.
- Hardening of tissue cultured planting materials.
- Increased carbon-dioxide levels with in greenhouses results in better photosynthesis.
- Plant growth is improved and crops mature faster.
- Use of water is optimized due to reduced evapotranspiration; up to 40-50% of water can be saved.

- Use of UV stabilized film filters out harmful ultra-violet rays.
- Cultivation in problematic agro climates and in problematic soil.
- Demand for fresh vegetables around metropolitan cities that sustains greenhouse production.

Constraints

- The cost of construction and operational cost of the climatic controlled green house is very high, which do not suit to the Indian growers without a strong government support.
- Regular power supply is required for operating cooling and heating system of the green house, which is not available almost in all parts of the country at farm level.
- Very little work on standardization of designs of greenhouses and other protected structures have been done for different agro-climatic regions of the country.
- Cladding material of required quality is not readily available in Indian markets and very costly also. There is also a lack of suitable instrumentation for environment control in green houses.
- Lack of indigenous information base on the relevant technologies for use by the prospective users and entrepreneurs.

Concepts and classification

Construction of greenhouses should be based on the following two principles.

1. Structures to provide cooling in hot areas

In hotter areas, cool environment has to be created inside the green house to reduce the temperature and high humidity by providing ventilators, cooling pad and fans, misting arrangements, foggers, covering with shade net etc.

2. Structures to provide heating in cool areas

In cooler areas, facilities for warming up the micro climate has to be created inside the green house by providing artificial heating units, humidifiers etc.

Classification of greenhouses

I. Based on cost

a) Low cost or low tech greenhouse

Low cost greenhouse is a simple structure constructed with locally available materials such as bamboo, timber etc. The ultra violet (UV) film is used as cladding materials. Unlike conventional or hi-tech greenhouses, no specific control device for regulating environmental parameters inside the greenhouse is provided. Simple techniques are, however, adopted for increasing or decreasing the temperature and humidity. Even light intensity can be reduced by incorporating shadenets. The temperature can be reduced during summer by opening the side walls. The temperature inside is increased when all sidewalls are covered with plastic film. This type of greenhouse is mainly suitable for cold climatic conditions.

b) Medium-tech greenhouse

Greenhouse users prefer manually or semi-automatic control type of structures due to minimum investment. This type of greenhouse is constructed using galvanized iron (G.I) pipes. The cladding materials on the roof are attached to the structure firmly with the help of screws. Whole structure is firmly fixed with the ground to withstand the disturbance against wind. Exhaust fans with thermostat are provided to control the temperature. Evaporative cooling pads and misting arrangements are also provided to maintain a favourable humidity inside the greenhouse. Hence, the system is semi-automatic, requires a lot of attention and care and it is very difficult to maintain uniform environment throughout the cropping period. These greenhouses are suitable for dry and composite climatic zones.

c) Hi-tech greenhouse

Hi- tech greenhouse is entirely automatic. All the environmental parameters are supported to function automatically.

Cost involved

1.	Less expensive greenhouse without fan and pad	Rs.300 to 500/m^2
2.	Medium cost greenhouse with pad and fan system without automation	Rs.800 to Rs.1100/ m^2
3.	Expensive greenhouses with fully automatic control system	Rs.2000 to Rs.3500/ m^2

The greenhouse can also be classified based on type of structures, type of glazing material, number of spans, environmental control etc. Various types are as follows:

1. Classification as per type of structure
    a) Quonset type
    b) Curved roof type
    c) Gable roof type

2. Classification as per glazing materials used
    a) Glass
    b) Fiberglass reinforced plastic materials
        i) Plain sheet
        ii) Corrugated sheet
    c) Plastic film
        i) Ultra violet stabilized low density polyethylene
        ii) Silpaulin
3. Classification based on number of spans
    a) Free standing or single span
    b) Multi-span or ridge and furrow or gutter connected
4. Classification based on environmental control
    a) Naturally ventilated
    b) Passive ventilation

Design of greenhouse

The design of greenhouse should be based upon sound scientific principles which facilitate controlled environment for the plant growth. Controlled environment for plant production systems are widely used throughout the world to produce plant materials and products at a time or place, or of a quality that cannot be obtained outdoors. Controlled environment agriculture requires more capital investment per unit area than field agriculture and thus must essentially be correspondingly more intensive to justify investment costs. Hence, the main components of greenhouses such as the basic structure, covering/glazing materials and temperature control systems need proper design for healthy growth of plants.

Under Indian conditions, Quonset type, multispan greenhouses are the most suitable, because of its low cost and ease of fabrication. Ultra violet resistant low-density polyethylene (UVLDPE) single film cladding of 200-micron thickness is sufficient for Naturally Ventilated (NV) greenhouses as well as Fan and Pad (FP) type of greenhouses. This should be fully tightened by stretching on the structure to avoid flutter and tearing. It should not be nailed or screwed to the structure as it gives the chance for tearing. The T-Lock or L- Lock should be used for fastening the sheet at structure, as this does not tear the sheet and sheet replacement is easy. Carbon dioxide enrichment can be done to improve the yield by maintaining its concentration at 1000 ppm.

Components of a green house

Roof: It is the top cover of a green house.

Gable: This is the transparent wall of a green house.

Cladding material: The transparent material mounted on the walls and roof of a green house.

Rigid cladding material: cladding material with such a degree of rigidity that any deformation of the structure may result in damage to it. e.g., Glass.

Flexible cladding material: It is the covering material with such a degree of flexibility that any deformation of the structure will not result in damage to it. e.g., Plastic film.

Gutter: It is a provision which collects and drains rain water and snow which is in place at an elevated level between two spans.

Column: It is the vertical structure supporting or carrying the greenhouse structure.

Purlin: It is the structure which connects cladding supporting bars to the columns.

Ridge: It is the highest horizontal section in top of the roof.

Girder: It is a horizontal structure, connecting columns on gutter height.

Bracings: These are given to support the structure against wind.

Arches: These are the structures supporting covering materials.

Foundation pipe: These are connections between the structure and ground.

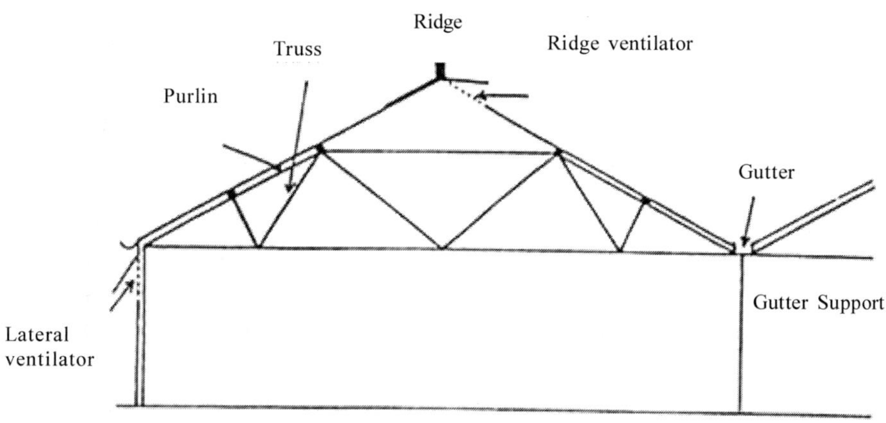

Components of typical greenhouse

Special training practices in green house cultivation

a) **Vertical training:** It is a commonly used training method, most probably used for almost all vegetables. The plants are allowed to reach the horizontal wires before being stopped.

b) **Arch:** Vertical training is frequently used in amateur circles on any system of short term culture. In long term crops, the plants can be arched over the wires that can be grown continuously.

c) **V-Training:** When the plants are grown so closely, can be trained alternatively in different directions i.e, 'V' shape. It is most probably used in straw bale culture.

d) **S-hook system:** The 16-gauge S-hooks are set 14 inch apart and the plants are supported as the season progresses, the plants are progressively layered. Wires or plant hooks are desirable to avoid the bottom truss trailing on the growing medium. At the end of the rows, the plants are turned on to the next row.

e) **Layering:** The plants are initially trained vertically, being completely detached when they reach the horizontal wire, and layered usually at the stage where lower trusses are picked and where the ripening trusses are kept well above growing medium level.

f) **Lateral training:** The plants are trained at an angle of 35-40° wide mesh or polypropylene netting as a useful supporting medium for accomplishing this technology. The plants are tied on to the netting with small loops of string or wire/paper clips. It is usual to lower the plants as the season progress.

a) Vertical Training

b) Arch

c) V- Training

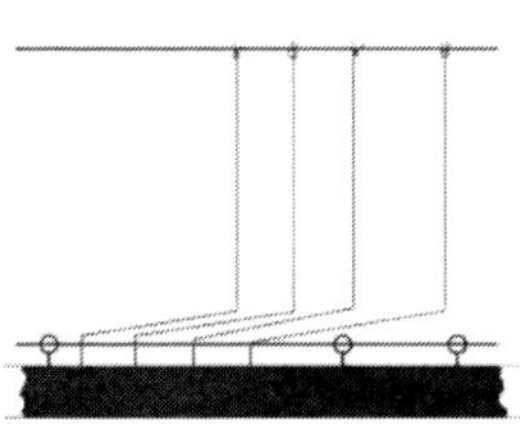

d) S- Hook system

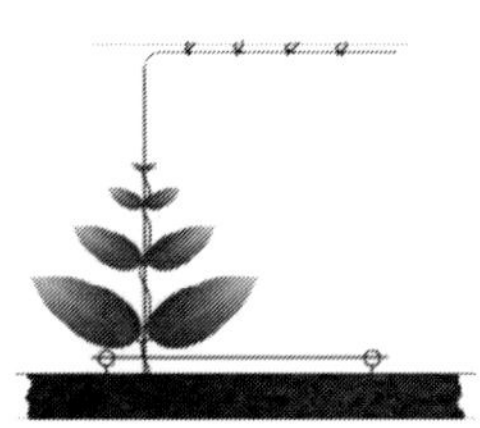

e) Layering

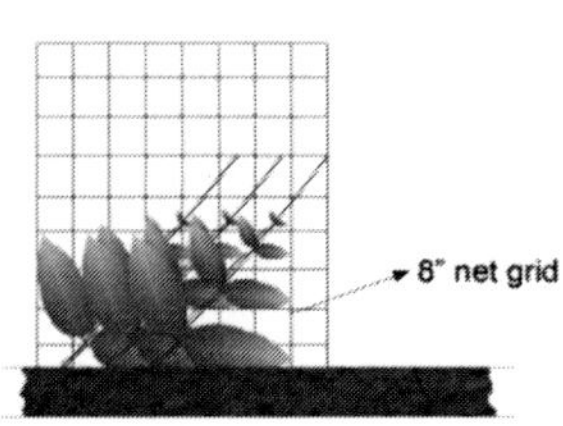

f) Lateral training

The choice of any particular training system will depend not only on the length of the growing season, but also on the type of greenhouse and whether ventilation can be effectively carried out to overcome the problem of disease which occurs more readily in systems other than the vertical. Laterally trained plants from the outset appear to be most prone to disease attack, due to air movement restriction, and plants layered later in the season eventually suffer a similar disadvantage.

Special cultural operations in greenhouse

1. Pruning

Pruning of lower leaves and all lateral shoots, allowing one single stem gives healthy shoots and leaves with productive flowers. In most of the vegetables, the old, pest and disease affected parts may be removed when grown under protected culture. In tomato, pruning of lateral shoots is commonly referred to as removal of "suckers". One sucker will form at the point where each leaf originates from the main stem, just above the leaf petiole. Allowing all suckers to grow and bear fruit would increase the total number of fruits, but they would be small and of poor quality. It is better to have one main stem that bears fruit, as this will produce larger, more uniform and higher quality fruit. The suckers should be removed once in a week to control the growth. Generally, any sucker longer than one inch should be removed. There are various pruning methods commonly followed in vegetables as follows:

i) **Side shoot removal:** This pruning system must be carried out by restricting the plants to one main stem. Removal may be done when the side shoots are young and preferably turgid. The little shoots may be snapped off with the finger and thumb. Suckers formed at ground level should be removed. When shoots are allowed to become too large, before removal, the scar left can be prone to diseases, especially later in the season.

ii) **Removal of lower leaves:** The matured useless leaves, which show yellowing, may be removed. These older lower leaves are the source of fungal inoculam. It is usual to remove the foliage, up to the ripening truss. One of the main benefits of leaf removal is that air is allowed to circulate through the base of the plants. In some cases, the lower leaves of up to 3 feet may be removed ruthlessly but it has certain physiological effects on the plant. Application of any one fungicide may be effective which prevent disease infection.

iii) **Stopping:** Stopping may be done usually when there is little hope that any further fruit trusses will form and will ripen before the plants are removed, either because the season is coming to a close or in the interests of another crop. Usually this is done up to the height of the top wire, unless the plants are to be arched or layered. The main reason for stopping is to avoid a cluster of leaves on the roof of the greenhouse.

2. CO_2 Enrichment

About 40 per cent of the dry matter of plants is composed of carbon. Under normal conditions, Carbon dioxide exists as a gas in the atmosphere slightly above 0.03 per cent or 345 ppm. During the day, when photosynthesis occurs under natural light, the plants in a greenhouse draw down the level of CO_2 to below 200 ppm. Under these circumstances, infiltration or ventilation increases CO_2 levels, when the outside air is brought in, to maintain the CO_2 at ambient levels. If the level of CO_2 is less than ambient levels, by providing ventilation may be uneconomical, due to the necessity of heating the incoming air in order to maintain proper growing temperatures.

CO_2 levels can be maintained economically even with 20% ventilation so that the crop response from ventilation is about 47 to 48% annually. The exact CO_2 level needed for a given crop will vary, since it must be correlated with other variables in greenhouse production such as light, temperature, nutrient levels, cultivar and degree of maturity. Most crops will respond favourably to CO_2 at 1000 to 1200 ppm.

3. Mulching

Mulching is the practice of covering the soil surface around the root zone of the plant with suitable materials to create congenial conditions for the growth and development of crop. A typical arrangement of plastic mulch application for crop production has the features like mulch film laying over the crop bed, drip irrigation lines under the film, holes on the film for plant growth, and the film secured by burying its edges. It is an effective practice under greenhouse cultivation as it restricts weed growth, reduce compaction, conserve moisture and reduce the effect of soil borne diseases through soil solarisation. Black plastic mulch is most commonly used under tropical conditions like India, but clear, white and silver grey colours are also used on certain occasions. Black plastic mulch along with drip system helps to reduce the soil temperature in tropical greenhouse cultivation, while transparent clear plastic mulch warms up soil the most. The clear plastic allows the sun rays to penetrate through the mulch directly and hence, warming the soil. The plastic helps to hold the heat. The disadvantage of clear plastic mulch is that the sunlight passing through the mulch also allows weeds to grow.

Saw tooth type of polyhouse

Shade net house

Hydroponics

Hydroponics is a technology for growing plants in nutrient solutions with or without the use of an artificial medium such as sand, gravel, vermiculite, rock wool, peat moss and saw dust to provide mechanical support. Liquid hydroponic systems have no other supporting medium for the plant roots, whereas aggregate systems have a solid medium for support. Hydroponic systems are further categorized as open system when the nutrient solution is delivered to the plant roots and not reused, and closed system when the surplus solution is recovered, replenished, and recycled.

Hydrponic systems

Aggregate system	Non-aggregate system
(Medium present) (or liquid system)	(Medium absent)
1. Gravel culture	1. Nutrient film technique (NFT)
2. Sand culture	2. Modified NFT
3. Peat culture	3. Floating hydroponics
4. Pumice culture	4. Aeroponics
5. Perlite culture	
6. Rockwool culture	
7. Smoked rice husk culture	
8. Saw dust culture	

Further classifications

1. Open System

Nutrient solution is delivered to the roots and not reused.

2. Closed System

Nutrient solution is delivered to the roots. Surplus solution is recovered, replenished and reused.

Based on the movement of solution, the hydroponic systems are classified as -

Active systems: In this system, the nutrient solution moves actively using a pump

Passive systems: In this system, the pump is not used. The plant takes up nutrient through a fibrous wick from the pot to the nutrient solution. This solution will help as a bubbler to maintain proper oxygenation.

The above systems are further classified based on the techniques method of culture.

1. Liquid hydroponics (solution culture or aqua culture)
   a) Circulating systems (closed systems)
      i) Nutrient film technique (NFT)
      ii) Deep flow technique (DFT)
   b) Non- circulating methods (open systems)
      i) Root dipping technique
      ii) Floating technique
2. Aggregate systems (solid media culture): These can be open systems or closed systems
   i) Hanging bag technique
   ii) Grow bag technique
   iii) Trench or trough technique
   iv) Pot technique
3. Aeroponics
   i) Root misting technique
   ii) Fog feed technique

Advantages of hydroponic systems

- It is one of the high density planting system accommodating more plants and giving higher output per unit area
- Maximum crop yield, earlier maturity and rapid turn over of crops.
- Crop production possible where no suitable soil exists.
- Freedom from the constraints of ambient temperatures and seasonality.
- More effective use of water and fertilizers.
- Minimal use of land area and also even uncultivable land can be brought in use.

- Suitable for mechanized production and plant protection.
- Isolation of the crop from the underlying soil which may have problems associated with nematodes, diseases, weeds, salinity, alkalinity, poor structure and drainage.
- Soil containing harmful microbes is not used and hence the produce is healthy, fresh and nutritious. There is no soil borne diseases and nematode damage.
- Pests and diseases are easier to get rid off than in soil because of the mobility of water.
- No hazards of continuous cropping.
- Production of crops during off-season and also throughout the year.
- Produce of vegetables grown in hydroponic system are free from puffiness (due to Ca and K deficiency) and therefore, possess better shelf life and stand long distance transport.

Disadvantages of hydroponics

- High costs of capital and energy inputs relative to conventional open-field agriculture (OFA).
- High degree of competence in plant science and engineering skills required for successful operations.
- Without soil as a buffer, any failure to the hydroponic system leads to rapid plant death.
- Pathogen attacks such as damping-off due to *Verticillium* wilt, is caused by the high moisture levels associated with hydroponics and over watering of soil based plants.
- Many hydroponic plants require different fertilizers and containment systems.
- To produce the mineral wool and the fertilizers that are needed to be used in this method, a large amount of energy is required.

Therefore, successful application of hydroponic technology is limited to:

- High value crops.
- Specific regions.
- Specific times of the year.
- When comparable open field crops are not readily available.

Components of hydroponic systems

A. Structures and environmental control

The structures and techniques used in controlled environment horticulture (CEH) are also used in the hydroponic systems. The cost of greenhouse (or) glasshouse structure decides the total cost. The greenhouse must admit the visible light portion of solar radiation and therefore, must be transparent. At the same time, to protect the plants, a greenhouse must be ventilated (or) cooled during the day because of the load from the radiation and it must be heated (or) insulated during cold nights. Technically, it would be possible to satisfy these contradictory criteria by growing plants in a windowless, insulated, heated (or) air-conditioned box using artificial illumination only.

Glasshouses are the main frames for the hydroponic system of plant cultivation. The solar energy could be trapped. Glasshouses are popular in NW Europe, especially in Netherlands, Belgium and England. Glasshouse (or) greenhouse structures are inevitable in the tropics where heavy winds may cause considerable damage.

B. Medium

The choice of medium depends on availability, cost, quality and type of hydroponic system to be used. The medium i) should possess high water retention capacity which is determined by particle size, shape and porosity, ii) should not contain any toxic substances, iii) should be hard enough to have long durability and iv) should be cheap and readily available. Examples are sand, gravel, saw, dust, peat, perlite, rockwool, pumice, smoked rice husk etc; Among these, the use of horticultural rockwool as the growing medium in open hydroponic system has been increasing rapidly. As a growing medium, rockwool is not only relatively inexpensive but also biologically non-degradable, takes up water easily, approximately 96% "pores" (or) interstitial air spaces, has evenly sized pores, ends itself to simplified and lower-cost drainage systems, and is easy to bottom heat during winter. It may be incorporated as a soil amendment after crops have been grown in it for several years.

C. Nutrition

All the essential elements are to be supplied to the plants through roots. The required fertilizer salts are to be dissolved in water to prepare a nutrient solution. The cost of the nutrient chemicals and their solubility are to be considered. Nutrient solutions have been standardized for different crops.

Knobs Solution

Ca NO_3 0.8 g/litre + β pinch of ferric solution is added

K NO_3 0.2 g/litre

Mg SO_4 0.2 g/litre

KH_2 PO_4 0.2 g/litre

Micronutrients (Hass and Reed's A to Z solution)

Boric acid 0.6 g/litre Mn Cl_2 0.4 g/litre

$ZnSO_4$ 0.05 g/litre Cu SO_4 0.05 g/litre

Al SO_4 0.05 g/litre KI 0.03 g/litre

KB 0.03 g/litre $CO(NO_3)_2$ 0.05 g/litre

HCI 0.03 g/litre Sn Cl_2 0.03 g/litre

Ni SO_4 0.05 g/litre Ti Oz 0.03 g/litre

Origin of essential elements in soil and hydroponics

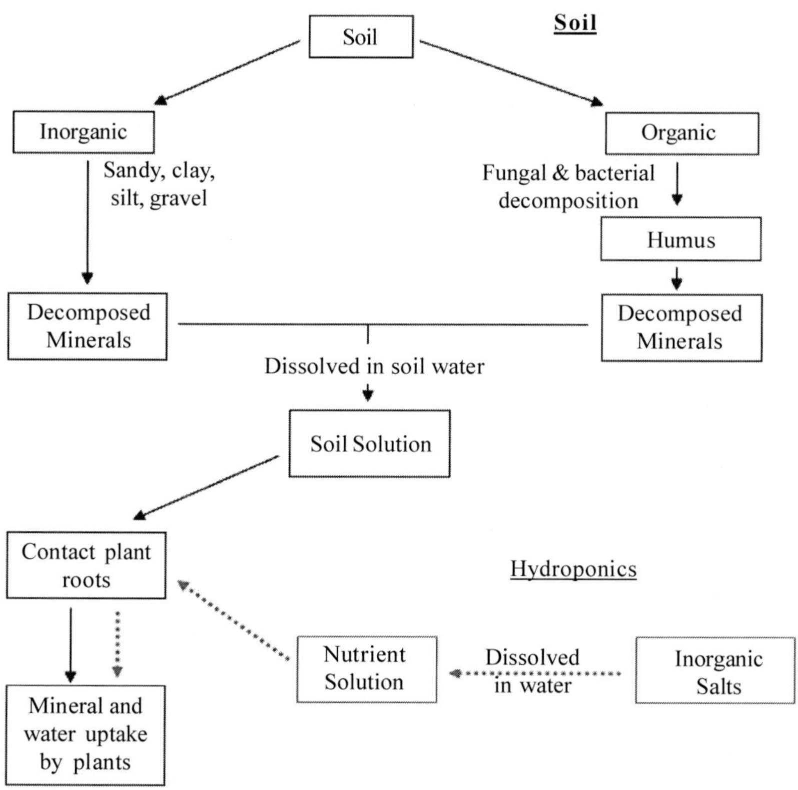

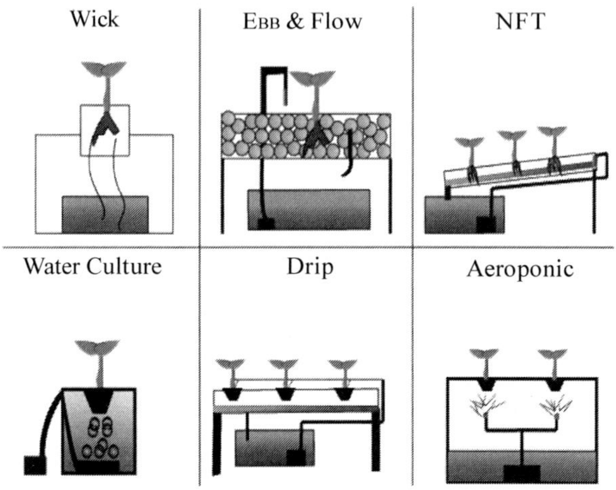

Nutrient Film Technique (NFT)

It is a technique for cultivation in water in which the plants grow with their root system within a sheet or opaque plastic, through which the nutrient solution is continuously circulated. The plastic forms tubes that are closed around each plant, and clipped together. They are usually polythene and lasts for up to 6 years. At the base of the plastic, below the plant, a thin sheet of rock wool is placed as a capillary mulch. This ensures that none of the plants dries out in the early phases of growth, while leaving most of the roots exposed, allowing good aeration. The roots grow freely and are mainly located at the bottom of the plastic tube, in direct contact with the solution.

The basic principle of NFT is the simple one of re-circulating a shallow stream of nutrient solution over the bare roots of growing plants to provide adequate water, nutrients and aeration. The plants are usually grown in a parallel series of sloping gullies lined with polyethylene sheets. The nutrient solution is pumped to the upper end of the gullies and flows around plant roots until it reaches a catchment pipe that returns the solution by gravity to a tank situated below ground level. The nutrient solution in this tank is repeatedly monitored (EC, temperature, pH) and the water and salts are replenished manually or automatically.

In a nutrient film system, a thin film of nutrient solution flows through plastic lined channels, which contain the plant roots. The walls of the channels are flexible, this permits them to be drawn together around the base of each plant, excluding light and preventing evaporation.

Nutrient solution is pumped to the higher end of each channel and flows by gravity to the plant roots to catchment pipes and a sump. The solution is monitored for replenishment of salts and water before it is recycled. Capillary material in the channel prevents young plants from drying out, and the roots soon grow into a tangled mat.

NFT Design

A principal advantage of the NFT system in comparison with others is that it requires much less nutrient solution. It is therefore, easier to heat the solution during winter months, to obtain optimum temperatures for root growth, and to cool it during hot summers in arid or tropical regions, thereby avoiding the bolting of lettuce and other undesirable plant responses. Reduced volumes are also easier to work with if it is necessary to treat the nutrient solution for disease control. Following are additional points important to the design of NFT systems for the tropics.

The channels should not be greater than 15-20 m in length. In a level green house, longer runs could restrict the height available for plant growth, since the slope of the channel usually has a drop of 1 in 50 to 1 in 75. Longer runs and/or channels, with fewer slopes, may accentuate problems of poor solution aeration. To assure good aeration, the nutrient solution could be introduced into channels at two or three points along the length. The flow of nutrient solution into each channel should be 2-3 litres per minute, depending on the oxygen content of the solution. The maximum temperature of the nutrient solution is 30°C. Temperature above 30°C will adversely affect the amount of dissolved oxygen in the solution. There should be approximately 5 ppm or more, especially in the nutrient solution flowing over the root mat in the channel.

In tropical regions, it is important that the channels be coloured white in order to mitigate the problems of heat building up from direct sunlight. Normally a white on black plastic is used for the channels. High air temperatures are common in the tropics. An NFT system may permit economical cooling of plant roots, avoiding the more expensive cooling of the entire green house aerial temperature.

Advantages of NFT

- Saving water.
- Precise control of the plant's nutrition, which results in better use of the fertilizer applied.
- Less pests, etc., as the crop is grown in an ideal environment.
- The problem of soil fatigue is eliminated, and the risk of the plant suffering as a result of water shortage declines.
- It reduces cultivation tasks.
- The abundant production of high quality produce.

- The NFT system eliminates the time, material and labour costs for pasteurization between crops.
- NFT has the potential for conserving water and nutrients.
- Recirculation of solution provides an excellent method for reducing nutrient and pesticide effluent from greenhouses.
- The main advantage of the NFT, is its potential for automation.
- Formation, testing and adjustment of nutrient solutions can be handled at a central point, and even these operations can be automated.
- The nutrient solutions are mechanically delivered to the crop.
- It is possible to alter the heat level of the nutrient solution by heating or cooling to suit to the plant requirement.
- Use of heavy root substrates and their handling is eliminated.

Disadvantages of NFT

- The cost of installation and maintenance is high.
- It is complex and requires technical knowledge.
- The plants are at risk due to fungal diseases.
- The shock of transplantation is worse in hydroponic cultivation than in the field, because all the soil has to be washed off the seedling's roots.

Aeroponics

It is a closed system hydroponics, where plants are grown in holes on panels of expanded polystyrene or other material. The plant roots are suspended in mid-air beneath the panel and are enclosed in spraying box. The box is sealed so that the roots are in darkness, which inhibits algal growth in the prevailing saturation humidity. A misting system sprays the nutrient solution over the roots periodically. The system is normally turned on for only a few seconds every 2 to 3 minutes. This is sufficient to keep roots moist and nutrient solution aerated. The A-frame aeroponic system developed in Arizona for leafy crops may be feasible for commercial food production. Inside a controlled environment structure, these frames are oriented with the inclined slope facing east-west.

The expanded plastic panels are of standard size (1.2 m x 2.4 m), mounted lengthwise, and spread 1.2 m at the base to form an equilateral triangle on the end view. The A-frame rests on the top of a panel-sized water-tight box, 25 cm deep, which contains the nutrient solution and misting equipment. Young seedlings in small cubes of growing medium are inserted into holes in the panels, which are spaced at intervals of 18 cm centre to centre. An apparent disadvantage of such a system is uneven growth resulting from variations in light intensity on the crops growing on the inclined panels.

Advantages of Aeroponics

- An advantage of this technique for lettuce or spinach production is that twice as many plants can be accommodated per unit of floor area than in other systems.
- Aeroponics is commercially successful for propagation, seed germination, seed potato production, tomato production and leaf crops production.
- Any species of plants can be grown because the micro environment of an aeroponics can be precisely controlled.
- Plants receive 100% of the available oxygen and carbon dioxide to the roots zone, stems, and leaves thus accelerating biomass growth and reducing rooting times.
- Can be done in uncultivable soils.

18

Good Agricultural Practices

GAP stands for "Good Agricultural Practices" which are used for production of safe agricultural produce which has zero microbial load on it and inside it. It is important for the producer to be acquainted with GAP.

- Good Agricultural Practice of horticultural crops are intended to apply to the growing and processing of all produce as traded and used in the European Union. Hence, they apply to the production of all plant materials used in the food, feed and medicinal industries. They also apply to all methods of production including organic production in accordance with the European Union regulations.
- The farmers involved in the cultivation of horticultural produce should ensure that the there is no damage to the existing wild life and the biodiversity in the farms which are maintained. It is advisable to regulate wild gathering by a specific guideline.
- The main aim of GAP guidelines is to ensure that the produce meets the demand of the consumer and is of high quality standard.
- It is therefore, important that they are produced hygienically, in order to reduce microbiological load to a minimum. They are produced with care, so that negative effect on plants in the course of cultivation, processing and storage is limited.
- All participants in the production process are required to comply with these guidelines voluntarily and to develop practical measures to achieve it.

Principles and guidelines for Good Agricultural Practice (GAP)

1. Seeds and propagation material

- Seeds in use are to be identified botanically, indicating plant variety, cultivar and origin. The material used should be 100% traceable.
- Vegetative propagated material also should be traceable so the information on parent material is important along with other vital information. Parent material used in organic production has to be certified as authentically organic.

- The parent material should meet the required standards of purity and germination. Efforts should be made to use certified seed as propagation material. The parent material should be free from pests and diseases in order to guarantee healthy plant growth. Preferrence should be given to resistant or tolerant species.
- The occurrence of plants and plant parts that are not similar to the species or variety planted, should be culled at all stages viz.; planting, growth, harvest, drying and packaging. Plant materials or seeds derived from or comprising genetically modified organisms have to comply with national and European Union regulations.

2. Cultivation

Depending on the method of cultivation e.g., conventional or organic growers should be allowed to follow different procedures for cultivation. The principles of good crop husbandry must be followed including an appropriate rotation of crops.

Soil and Fertilization

- Medicinal and aromatic plants should not be grown in soils that are contaminated by sludge. Furthermore, soils should not be contaminated by heavy metals, residues of plant protection chemical and any other unnatural chemicals.
- The manure applied should be devoid of human faeces and be thoroughly composted prior to application.
- All other fertilizing agents should be applied sparingly and in accordance with the demands of the plant and the particular species (including application between harvests). The use of fertilizers should be in accordance with efforts to minimize leaching.

Irrigation

- Irrigation should be minimized as much as possible and only be applied according to the needs of the plant.
- Water used for irrigation should comply with national and European Union quality standards and be as free as possible from contaminants, such as faeces, heavy metals, pesticides, herbicides and toxicologically hazardous substances.

Crop maintenance and plant protection

- Tillage should be adapted to the growth and requirements of plants.

- Pesticide and herbicide application should be avoided as far as possible. Products for chemical plant protection have to conform to the European Union's maximum residue limits (European Pharmacopoeia, European Directives, Codex Alimentarius). Application and storage of plant protection products have to be in conformity with the regulations of manufacturers and the respective national authorities.
- The application should only be carried out by qualified person, using approved equipment. Application should precede the harvest by a period either defined by the buyer or indicated by the producer of the plant protection product.
- The use of pesticides and herbicides has to be documented.
- All procedures regarding nutrient supply and chemical plant protection, should ensure safe marketability of the product.
- The buyer of the produce should be given in writing, about the chemical brand, quantity and date of use.

3. Harvest

- Harvesting should take place when the plants are of best possible quality, according to their different utilizations.
- Harvesting should preferably takes place under the best possible conditions (wet soils, dew, rain or exceptionally high air humidity can be unfavourable). If harvesting is done under wet conditions, extra care should be taken in order to avoid the unfavourable influence of moisture.
- Equipment should be serviced time to time and also be kept in clean. Those machine parts, including their housing, that have direct contact with the harvested crop should be regularly cleaned.
- Cutting devices of harvesters must be adjusted so it collects the least soil debris.
- While harvesting, it should be ensured that no weeds are mixed with the harvested crop, if so, it should immediately be removed.
- All containers used in the harvest must be cleaned and kept free of the remnants of previous crops; containers not in use must also be preserved in a dry condition, free of pests and inaccessible to mice/rodents as well as livestock and other domestic animals.
- The harvested crop should not come into direct contact with the soil. It must be promptly collected and put in transport containers.
- Mechanical damage and compacting of the crop that may result in undesirable quality changes must be avoided. Attention must be paid in ensuring that the

sacks are not overfilled and that the stacking up of sacks does not result in compacting of the produce.

- The produce should be transported safely, without causing any heating of the produce.
- Delivery of freshly harvested plant material to the processing facility must occur as quickly as possible in order to prevent it's heating up.
- The harvested crop must be protected from pests, mice/rodents and domestic animals. Pest control measures should be documented.

4. Primary processing

- On arrival to the processing unit, the harvested produce should be promptly unloaded and unpacked safely, without exposing it to any detrimental conditions like harsh sun or rain.
- Buildings used in the processing of harvested crops must be clean and should have proper ventilation. It should be separate and not be used for any other purpose.
- Buildings which store the harvested produce should be protected against birds, insects, rodents and animals.
- Processing equipment must be kept clean and be regularly serviced.
- In the case of natural open air drying, the crop must be spread out in a thin layer to avoid mould formation and to acquire uniformly dried product.
- The temperature, duration of exposure, etc. must be selected according to the crop, so that the quality of the produce is maintained.
- All material must be sorted and graded to maintain the highest quality standards.
- Waste-bins should be kept handy, emptied and cleaned daily.
- The dried product should be promptly packed and stored in required conditions.

5. Packaging

- After the repeated control and eventual elimination of low-quality materials and any foreign bodies, the product should be preferably packaged in new, clean and dry sacks, bags or chests. The label must be clear, permanently fixed and be made of non-toxic material. Information must conform to the European Union and national labelling regulations.
- Packaging materials should be stored in a clean and dry place that is free of pests and inaccessible to livestock and other domestic animals. It must be

guaranteed that no product contamination occurs as a result of using packaging material, particularly in the case of fiber bags.

- Reusable packaging materials should be well cleaned and perfectly dried prior to their usage. It must be guaranteed that no contamination takes place by reusing bags.

6. Storage and transport

- Packaged dried materials and essential oils should be stored in a dry, well aerated building, in which the daily temperature fluctuations are limited and good aeration is guaranteed.
- Frozen products should be stored below-18°C (or below-20°C for longer term storage).
- Essential oil storage must conform to the appropriate chemical storage standards.
- As a protection against pests, birds, rodents and domestic animals, the window and door openings should be protected, e.g. by wire netting.
- It is recommended that the packaged dry crop be stored in buildings with concrete or similar easily cleanable floors, on pallets maintaining sufficient distance from the wall, with thorough separation from other crops to avoid cross-contamination and organic products must be stored separately.
- In case of bulk delivery, it is important to ensure that the transportation conditions are good. Transportation containers should be such that the produce quality is not altered.
- Fumigation against pest attack should only be carried out on necessity and must exclusively be administered by licensed personnel. Only registered chemicals should be used. Any fumigation against pest attack should be reported in the documentation.

7. Personnel and facilities

- People involved in the production chain should be educated on the procedures, the risks involved and the safety measures.
- All processing procedures should fully conform to both EU-guidelines on food hygiene and the general food hygienic principles of Codex Alimentarius as well as the European directive on Good Manufacturing Practice (GMP).
- The people involved in the production should not be allowed in the production place when they are infected and infested.They should be trained in personnel hygiene. Safety gears should be issued to them and they should be trained to wear the safety gears whenever a production procedure warrants it.

- Toilets, hand washing facility and eating area should be provided and should meet the regulatory standards.
- Persons suffering from known infectious diseases transmittable via food, including diarrhoea, or being transmitters of such diseases, must be denied entry into areas where they may come into contact with the plant material, as per relevant regulations.
- Persons with open wounds, inflammations and skin infections should be kept away from those areas where plant processing takes place, or have to wear appropriate protective clothing or gloves, until they recover completely.
- The welfare of all staff involved in the production chain should be ensured.

8. Equipment

- Equipment used in plant cultivation and processing should be easy to clean, in order to eliminate the risk of contamination.
- All machinery should be mounted in an easily accessible position. They must be well serviced and cleaned regularly. Fertilizer and pesticide application machinery must be regularly calibrated.
- Equipment should be of the material which does not react with the product and deteriorate the product quality.

9. Education

It is highly advisable to educate all persons who are involved in the production chain. They should be taught about the safe cultural practices, including safe ways to use insecticides, fungicide, pesticides and herbicides. They should be told about the minimum days after which a sprayed produce can be harvested, so that it has no chemical residue.

10. Documentation

- Details of the field, geographical location, soil profile, previous crop in the field, the cultivation procedures of the previous crop should be documented.
- All parent materials, the source, treatment of parent material and planting procedures should be documented.
- Quality of inputs like water, fertilizers and chemicals should be documented.
- Field records of steps of cultivation, incidence of pests and diseases and the chemicals used thereof should be documented including the number of times a particular chemical was used, with dates.
- The stage of harvesting, the date of harvesting and the harvesting procedures should be recorded.

- If batches of produce are mixed from different areas, it should be documented. Similar crop of same quality may be mixed, but details should be documented.
- It is essential to document the type, quantity and date of harvest of the crop, as well as the chemicals and other substances (e.g. fertilizers, pesticides and herbicides, growth regulators, etc.) used during production.
- All procedures that affect the product quality must be entered into the batch documentation.
- In produce buy back contract, all agreements between the producer and the buyer, regarding production procedure and handling of produce should be written down clearly.
- Audit results should be documented and should be stored for a minimum of 10 years.
- All batches of crop should be labelled, giving details of the location, cultivated area, date of sowing and harvesting.
- Any special circumstances during the growth period and harvest of the crop, which may influence the chemical composition of the product, e.g. drought, floods, pests or disease outbreak must be documented.

11. Quality guarantee

The buyers can ask questions about the quality of the produce like when it was sprayed, with what it was sprayed etc. In case of medicinal and aromatic plants, the buyers can question the quality parameters like active principles, plant protection chemical residues etc. A written guarantee of the produce should be given by the producer. This should state the authenticity of the produce in terms of variety, chemical residue and purity.

19

Organic Horticulture

Organic cultivation is a deep rooted cultivation tradition, which had been practiced in many parts of the world, before the advent of the practice of use of chemicals.

After the green revolution there has been exponential usage of chemicals in agriculture, right from planting to harvesting. Chemicals are used in the form of nutrient inputs or for plant protection against pests, diseases, weeds and keeping deficiencies at bay. Nevertheless, after the awareness regarding the detrimental effect of chemicals on soil, climate and human health, there has been a lot of contemplation on substituting the chemicals with the organic alternatives. Growing crops using organic inputs is known as 'organic cultivation'. Organic cultivation is a sustainable form of cultivation where a proper balance is maintained and management of plant nutrition, pests, weeds and diseases is done in a sustainable manner, keeping the soil microflora intact and active. It does not release any chemical gases in the atmosphere, does not make the soil acidic and the soil does not lose its structure and texture. Contrary to chemical farming, where over a time period, the soil quality, texture and structure is lost, in organic practices, the soil texture, quality and structure improves over time, the plants become strong and the quality of the produce is improved. Other than organic inputs, organic farming relies on some sustainable practices like using native germplasm, crop rotations with legumes, green manuring, mulching, inter cropping, mixed cropping, incorporation of farm and animal waste compost.These simple age old practices build to organic content in the soil and elevate the soil's efficiency to make nutrients and water to the plants. The soil microflora and earthworms become biologically active, efficient and exponentially proliferate. The goal of organic cultivation is to get maximum produce from the soil in such a manner that the soil, air and water are clean and protected.

Advantages of organic Horticulture

- Maintenance of soil health.
- Prevention of soil degradation.
- Conservation of environment.
- Utilization of farm waste.

- Better quality produce.
- Premium price.
- A variety of products are available due to intercropping and mixed farming. A diversified organic farming system increases farm production by 20-30%.
- Since focus is on using native cultivars, the need for plant protection and nutrient inputs is minimised. These cultivars are resilient, thus surviving all kinds of vagaries.
- Can bring back the fertility of the land which has been rendered as dessert and degraded by intensive chemical based monoculture cropping system.
- Economic valorisation of native under-utilized cultivars and endangered germplasm.
- Conservation of biodiversity is feasible through organic farming.
- Food and shelter for birds, bees and butterflies is abundant in multicropping and multi-tier system without the presence of poisonous pesticides.

Aims of organic cultivation

- A sustainable horticulture system which improves soil fertility to ensure adequate food production in future without compromising on quality of life.
- A self-sustaining system which relies on inputs within the farm rather than outside the farm.
- A system which improves the soil microbes, quality, texture and structure, without causing an adverse effect on the environment.
- To get good quality produce.

Principals of organic farming

- Building soil fertility to adequate levels where the crop can take the required nutrients as per its requirement and the nutrients are released slowly.
- Recycling farm waste and compost into the farm.
- Management of pests, diseases and weeds sustainably by use of biocontrol agents, bio pesticides, and various cultural techniques like crop rotation, multiple cropping, mulching, crop rotation and green manuring.

Inputs of organic cultivation

Various components of organic cultivation are –

- Indigenous cultivars and breeds.
- Livestock and human waste.
- Plant and aquatic waste.

- Urban and rural waste.
- Crop residues.
- Green leaf manure and green manures like *Crotolaria juncea* (Sunhemp), *Sesbania aculeate* (Daincha).
- Cover crops-leguminous cover crops suppress weeds, prevents soil erosion, reduces evaporation of moisture from soil surface and fix nitrogen. It does not compete with the main crop for nutrients and water. It makes the nitrogen available to the main crops.
- Agro industry by-products and waste like coir pith, cocopeat.
- Biofertilizers for example *Rhizobium, Azospirillum,* AM fungi, *Azotobacter, Phosphobacteria* etc.
- Vermicompost.
- Panchakaviya.
- Neem oil, garlic oil, chilli powder, casuarina extract etc.
- Predators against pests.
- Trap crops for predators.

Method of application of organic inputs

- Composted manures, farm yard manure, crop residues, green manures are incorporated during land preparation. They are ploughed into the soil. Crop planting is done 2-3 weeks later. The compost and manure should not be left to dry in the field since the beneficial microbes are lost due to direct heat and sunshine.
- The biofertilizers may be applied as top dressing.
- Bioinoculants and biopesticides should be used in the recommended concentration and frequency.

Average composition of different organic manures

Organic Manure	N (%)	P (%)	K (%)
FYM	0.50-1.5	0.20	0.50
Sheep and goat manure	3.00	1.00	2.00
Poultry manure	3.00	2.60	1.40
Fish meal	4-10	3.90	0.30-1.50
Urban Waste	1.50	1.00	1.50
Sludge	1.50-3.50	0.75-4.00	0.30-6.00
Vermi compost	2.14	3.44	1.01
Neem cake	5.20	1.00	1.50
Groundnut cake	7.30	1.50	1.30
Sunhemp	2.30	0.50	2.20

Variations in organic cultivation

Integrated farming system

Low input organic farming, wherein the inputs available in the farm are effectively recycled. This type of farming is not just horticulture production, but also involves poultry farming, aqua culture, mushroom cultivation, goat rearing and dairy farming.

Integrated green manuring

High input technology which involves integrated nutrient management (INM) and Integrated pest management (IPM). Biofertilizers form an integral part of this kind of cultivation.

Pure organic farming

This kind of organic cultivation involves the use of organic manures, bio fertilizers, bioinoculants, biopesticides and special cultural practices. There is a total exclusion of chemical fertilizers and chemical inputs for pests, diseases and weeds.

Constraints in organic Horticulture

- It is labour intensive. In a multi cropping system, each crop has different schedule of intercultural operation. Even harvestings are separately done mostly manually, unlike monoculture where the machineries can easily be used, making the chemical monocropping systems less labour intensive.
- Availability of the organic inputs is a major issue.
- The organic inputs are bulky and difficult to transport.

Organic certification

Codex and IFOAM (International basic standards for organic production and processing) have laid out clear norms for organic produce for all of the following aspects –

- Field selection – The field selected for organic cultivation has some definite criteria. The field should be devoid of any kind of chemical spray for three years before taking the first organic harvest. The selected field should be sufficiently away from inorganic fields, so that there is no chemical contamination through ground water or drift.
- Production practices- No chemical should be used in disease, pest and weed management.

- Handling.
- Packing and storage-packing material used should be biodegradable and environment friendly.
- Transportation.
- Processing.
- Labelling.
- Marketing.

Codex and IFOAM strive to –

- Protect consumers against frauds.
- Ensure compliance of standards in all the processes from production to marketing.
- Give 'organic' label.
- Facilitate agreements for export.

20

Home Gardening

A kitchen garden/ nutrition garden is a vegetable garden where the vegetable crops are grown in the backyard of a house or any available space in compound or rooftop of the house, to meet the daily requirement of the family. All the members of the family are collectively involved in a kitchen garden.

For a balanced diet, 300 g of vegetables are to be included in our daily food. But on an average, Indians take only 160 g per day. Hence to fill up the gap, vegetable cultivation not only in farmers' holdings but also in the home gardens is encouraged.

Advantages

1. Growing vegetables by the family members serves as a good hobby and helps to save money in purchase of fresh vegetables.
2. The household gets pesticide residue free, clean vegetables, which is a rare commodity these days.
3. It helps to grow vegetables of our choice and we can consume it in fresh form.
4. Kitchen waste can be composted and used to grow organic and clean vegetables.
5. Waste water and land available within our house compound are best utilized for growing vegetables.
6. Kitchen waste can be composted and used in the kitchen garden.
7. It creates a healthy, beautiful atmosphere in the house as there is a common interest.
8. When all the members are involved, it gives them quality time together. The older people are not neglected.
9. It teaches dignity of labour to the young generation.

Due to increased cost, problem of pesticide residue and non-availability of fresh vegetables, every home should lay out a kitchen/ nutrition garden with available area. Arrangements may also be made to grow vegetables in containers and also on roof top, if little or no land is available.

Components of a Kitchen/ Nutrition Garden

The main aim of the kitchen garden is year round continuous supply of all kinds of vegetables needed for a household. The following components should be included in a kitchen garden.

Fence

It is very important to fence the garden to protect it from animals and trespassers. If no compound wall is provided, live fence can be grown. Bamboo thatties, barbed wire or plain wire can be used for fencing. For effective space utilization, coccinia, bitter gourd, lab lab and basella can be grown on the fence line.

Paths and irrigation channels

A main path dividing the entire garden into two halves with side paths and walks is made. The area for main and side paths should be the minimum. The width of the path should be 45 to 60 cm. The number of irrigation channels should also be at the minimum. Along the main path, pandal may be provided and above that grapes or any gourds can be trained. Along the side paths, greens like ponnanganni, palak, mint and small onions can be grown.

Perennial plot

This area should be located at the rear end of the garden so that the perennial plants do not affect the growth of other crops. Crops like moringa, curry leaf, tapioca, yams, agathi and fruits like lime, banana, west Indian Cherry can be grown in this area. The plants in this area are not disturbed every year. This is as good as a permanent feature of the kitchen garden.

Manure pits

Manure pits are dug at two corners of the garden at the rear end near the perennial plot. In this pit, garden and kitchen wastes including ash and household sweepings are dumped and composted. This can be used for manuring. Two pits are made so that when one pit is full, it can be left for composting and the other can be used to dump the waste. When second pit is full, the first one can be emptied and the compost can be utilized in the kitchen garden. Gourds like snake gourd, ribbed gourd can be grown near the manure pit and trained on pandal erected above the manure pit, so that the place above the compost pit is utilized.

Beds

After allocating areas for the above features, the rest of the area can be divided into rectangular beds of equal size. According to the area available, 6-8 beds may be formed. The size of the beds should be kept manageable, so that proper weeding and plant inspection is possible. On the ridges that separate the beds, radish, carrot, turnip or small onion may be grown, for effective utilization of the land area under cultivation.

Selection of vegetables

The selection of vegetable to be grown in a home garden depends upon the season, weather, choice of the family members and the size of the garden. Preference should be given to leafy and seasoning vegetables (spinach amaranth, lettuce, methi, onion, garlic, turmeric, ginger and coriander), as they grow faster. Crops which can be harvested continuously in the season for example tomato, chillies, brinjal, okra, peas, beans, lab lab etc. should be preferred. Cabbage, broccoli, brussels sprouts or cauliflower give a single harvest. Vegetables having good nutritive value and early maturity may also be selected. Some people may like to grow speciality vegetables like, broccoli, celery, parsley, endive, chive, patchouli, basil, asparagus, artichoke, red cabbage, coloured (red and yellow) sweet pepper and small pickling cucumber. Vegetables requiring large space for growing (melons, watermelon, pumpkin, etc.), are not suitable for growing in a home garden.

S.No.	Types of vegetables	Name of vegetables	Duration
1.	Fruit vegetables	Tomato, brinjal, chillies	4-5 months
2.	Root vegetables	Radish, carrot, beet root, turnip	3 months
3.	Bulb crops	Small onions, bellary onion, garlic	3-4 months
4.	Legumes	Lab lab, cowpea, french beans, peas, cluster beans	3-4 months
5.	Cucurbits	Pumpkin, bitter gourd, ribbed gourd, snake gourd, coccinia	4-5 months
6.	Tuber crops	Sweet potato, tapioca, yams	8-9 months
7.	Cole crops	Cabbage, cauliflower, knol-khol	3-4 months
8.	Leafy vegetables	Amaranthus, coriander, fenugreek, palak, Alternenthra, mint etc.	2 months

Crop rotation should be followed in each bed to retain the nutrient status of the soil and to ensure better performance of crops. Crop rotation may be done as follows:

- Shallow rooted vegetable (onion) should be rotated with deep-rooted one (lab lab and brinjal).

- Leguminous vegetable (cow pea) should be rotated with a non-leguminous vegetable (brinjal and tomato).
- Tuber-forming vegetable (sweet potato) should be rotated with non-tuber forming vegetable (bhendi, brinjal, etc.) or legumes, since tubers forming vegetables are heavy feeder. If rotated with legumes, the soil will be replenished with nutrients.

For continuous supply of vegetables almost throughout the year, green leafy vegetables may be sown or planted at different dates, preferably short duration crops first and later the long duration crop, so as to ensure regular supply of vegetables.

In general, vegetables could be grown throughout the year with a few exceptions. Cropping intensity should be the maximum.

A cropping programme is furnished for guidance.

Bed No.	Vegetables	Season of growing
1.	Brinjal + radish	June – September
	Cabbage	October – January
	Bhendi	February – May
2.	Tomato + cluster beans	June – September
	Beet root	October – December
	Greens	January – February
	Cowpea	March – May
3.	Lab Lab	June – September
	Brinjal and turnip	October – January
	Cluster beans	February – May
4.	Bhendi	June – September
	Cauliflower	October – January
	Radish	February – March
	Greens	April – May
5.	Chillies + onion (Small)	June – November
	Greens	December – January
	Brinjal + radish	February – May
6.	Onion (Bellary)	June – September
	Chillies	October – February
	French beans	March – May

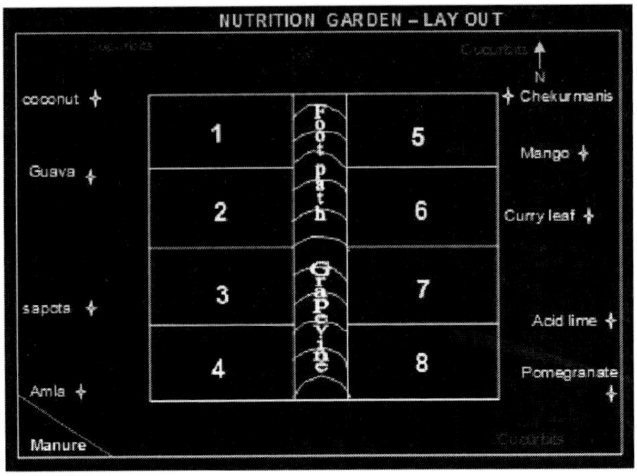

Crops and the duration

Crop	Duration		
Tomato	3 ½ - 4 months	Curry leaf	Perennial
Brinjal	4 months	Chekurmanis	Perennial
Chillies	5 months	Tapioca	Perennial
Moringa	Perennial	*Amorphophallus*	Perennial
Banana	Perennial	*Dioscorea*	Perennial
West Indian Cherry	Perennial	Colocasia	Perennial
Lime	Perennial	Spinach	Perennial

Cultural practices

Vegetables can be grown by seeds (okra, root crops, peas, beans, amaranth, palak, methi, coriander, etc.) or by transplanting seedlings (tomato, eggplant, chillies, sweet pepper, onion, cabbage, cauliflower, broccoli, knol-khol, etc.). Parval (pointed ground) is raised by planting rooted cuttings, potatoes by tubers, multiplier onions by small bulbs and garlic by cloves. In a limited garden space it may be useful to follow multiple cropping systems like intercropping, mixed cropping, relay cropping and succession cropping.

In a home garden use of chemical fertilizer and pesticides should be avoided or used judiously. Use of manures like farmyard manure, compost, cowdung manure, oil cake, vermicompost and green manuring are safe to human health and environment-friendly. Vermiculture may be adopted for producing vermicompost utilizing kitchen wastes. Nowadays organic farming practices and integrated pest management are being advocated for home gardening.

Generally furrow irrigation is practiced to supply water to plants. Water from tube well or bore well is used for irrigation. Drinking water from house taps should not be used for irrigation. Now-a-days harvesting of rain water is being encouraged in urban areas. Rain water collected in large plastic tanks can be used for irrigation.

Care for growing plants in nursery for transplanting

- Seedling rate may vary according to the soil temperature and probable germination percentage.
- Avoid too thick seeding, thus preventing damping off and spindly seedlings.
- Planting depth should be roughly 1-2 cm.
- Maintain the soil moisture in the desirable range for that particular crop seed.
- Seed beds should be free from weeds.
- Pre sowing soil disinfection or sterilization may help to avoid infestation with insects and diseases
- Spread seeds in a 'U' shaped furrow to get stocky seedlings.
- A stand of about 20-25 plants per metre of row will result in stockier plants than if larger population is attempted.
- Control insects and diseases carefully from the time of seedling emergence by prophylactic measure.
- Do not try to force plants too rapidly either by heavy fertilization, over watering or with higher temperature.
- Harden the plants before lifting by gradual reduction of watering for about 7-10 days.
- Soak the beds 6-12 hours before lifting to facilitate the maximum retention of roots and plant turgidity.
- Avoid shading or protection of seedlings; allow the maximum sunlight, except in case of expected injury due to heavy rain or frost.
- Dipping seedlings, just before transplanting, in the protection formulations may avoid pest or disease attack in the initial stages of establishment.

Pest and disease management

Pests

There are two types of pests, which normally attack the nursery plants.

1. Sucking pests – Aphids, white flies, thrips etc.
2. Biting (or) chewing pests-Beetles, grasshoppers, leaf eating caterpillars etc.

Apart from causing damage to the seedlings, the sucking pests also act as vectors for transmitting some of the viral diseases even in the nursery stage itself.

Eg. 1) Aphids spread mosaic disease in chillies, 2) Thrips act as vector for leaf curl virus disease in chillies and spotted wilt virus (tospo virus) disease in tomato.

Control

i) Application of systemic insecticides like Methyl demeton or Dimethoate @ one ml per litre of water by using a hand operated sprayer.

ii) Application of carbofuran @ 10g/sq.m 10 days before pulling of seedlings will also control the sucking pests in the nursery and at the early stages in the mainfield.

Diseases

Damping off (*Pythium* spp., *Phytophthora*, spp. *Rizoctonia* spp.)

Seedlings of tomato, chillies, brinjal, cabbage, and cauliflower are highly affected by this disease. Water logging with poor drainage leads to infection. The disease affects the seedlings in two ways.

1. Affected seeds get decayed inside the soil resulting in failure of germination.
2. After germination, the fungi attack the seedlings in the collar region. The tissues will become soft and succulent. In advanced stages, the seedlings become lodged and get decayed.

Control

1. Raising seedlings in raised beds with good drainage facility reduces the infection.
2. Addition of organic matter improves the soil texture and soil aeration.
3. Avoiding thick sowing of seeds in beds.
4. Periodical changing of nursery sites.
5. Treating the seeds with captan, thiram @ 2g or carbendazim @ 1 g/kg or *Trichoderma viride* @ 4g of seeds before sowing can reduce the infection.
6. Drenching the nursery bed once or twice with Bordeaux mixture 1% or copper oxychloride 0.3% can also reduce the infection.

Nematodes

Root knot and lesion nematodes commonly infect the seedlings. Before sowing the seeds, carbofuran @ 10 g/sq.m should be incorporated in the soil and watered regularly.

Ants: Application of Lindane 1.3% dust at the rate of 100 g/bed on all sides to protect the seeds from ants.

Herbal garden

Herbs are plants with aromatic properties, which are utilized for flavouring and garnishing foods. Herbs have medicinal properties and provide fragrances. Use of herbs is an important practice in Ayurveda and Unani systems of medicine in India and Chinese herbal medicine system. Common herbs are basil, thyme, rosemary, coriander, oregano, chive, parsley, lemon grass, vetiver, dill, cinnamon sage, lavender, marjoram, mint, pepper mint, chamomile, catnip, caraway etc.

Herbal gardens consist of commonly available and frequently used medicinal plants. The aim of herbal gardens is to conserving Rare, Endangered and Threatened (RET) species of medicinal Plants. National Medicinal Plant Board (NMPB) plays a vital role in India in promoting and conserving various types of herbs and setting various types of herbal gardens.

Types of herbs

Culinary herbs

These herbs are used for flavouring. Coriander, dill, mint etc. fall in this category.

Herbs for tea

These herbs can be infused to make herbal teas. They can be used as fresh, dried or in the form of seeds. Chamomile tea and mint teas are very popular.

Medicinal herbs

Herbs have been used to treat ailments since time immemorial in various ancient and indigenous systems of medicines around the globe. St. John's wort, kava, periwinkle, cannabis are some examples.

Sacred herbs

Turmeric, basil and cannabis are considered sacred herbs in India and are used in many religious rituals.

Herbs for cosmetics

Herbs are used to make fragrances, face cream, face pack and many more grooming things. Vetiver oil is the base for most of the perfumes. *Aloe vera* is used as rehydrating agent in facial creams, to mention a few.

Types of herbal garden

Herbal gardens can be grown at home, schools, institutes or of state level.

1. Home herbal gardens

Herbal gardens can be laid at home, in available space or in pots. The herbs that are used to treat the common every day ailments can be grown example basil, mint, ashwagandha, *Aloe vera,* periwinkle, rosemary, thyme, vetiver, etc.

2. School herbal garden

Setting up of herbal gardens in schools is a good way of teaching the young children about the commonly available and frequently used herbs and medicinal plants.

3. Institutional / Public herbal gardens

Herbal gardens in colleges like AYUSH Professional College students, universities, hospitals, etc. can help sensitize about the vast potential of indigenous knowledge of herbs which are present in India as an age-old practice in Ayurveda.

4. Herbal gardens of State and National Importance

Herbal gardens of national importance and state importance in various eco-regions of the country are planted to promote and popularize medicinal plants in an organized manner on a bigger scale.

Roof garden

Due to escalating population more land area is being brought under construction of houses, therefore, there is hardly any space for growing vegetables. Especially in multi-storeyed buildings, the only way to grow vegetables is in pots and containers. This practice is known as container gardening.

Aims of roof gardening

- To utilize the available space
- To utilize the available time in constructive hobby
- To grow organic vegetables free from pesticide and insecticide residues
- To grow our own favourite vegetables
- Rare and unavailable vegetables can be grown

Location of garden

Since there is not enough place in the flats these days, the garden can be laid on roof tops, thereby effectively utilizing the available space on the roof tops

In multi-storeyed buildings, not all the apartments have a roof. Thus the pots can be placed in the verandah and window sills.

Requirements for a roof garden

Types of containers

- Cement pots
- Earthen pots
- Plastic barrels
- Wooden barrels
- Boxes
- Crates
- Pans
- Plastic jars
- Buckets
- Tin boxes
- Drums

Requirement of tools

Hand hoe, spade, shovel, rose can, hand sprayer, gardening hose with sprinkler, bamboo stakes and jute strings are required.

Requirement of inputs

The inputs required for roof garden are quality seeds and planting materials from reputed institutes and firms, good soil free from stones, weeds and other undecomposable materials, well decomposed organic manures, river bed sand, chemical fertilizers, insecticides and fungicides.

How to start roof garden

Wash the container thoroughly and make drainage holes at the bottom. Mix soil, compost and sand in equal proportions with the help of the hand hoe and shovel. Fill the containers loosely and gently tap. The soil should settle, bearing one inch head space at the top for irrigation.

Select the container (grow bag or plastic or cement or mud pot)

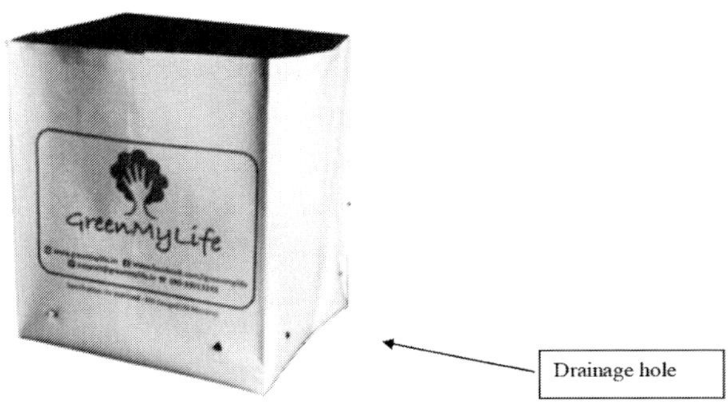

Make holes for draining the excess water

Then fill the medium in the pots & press it lightly

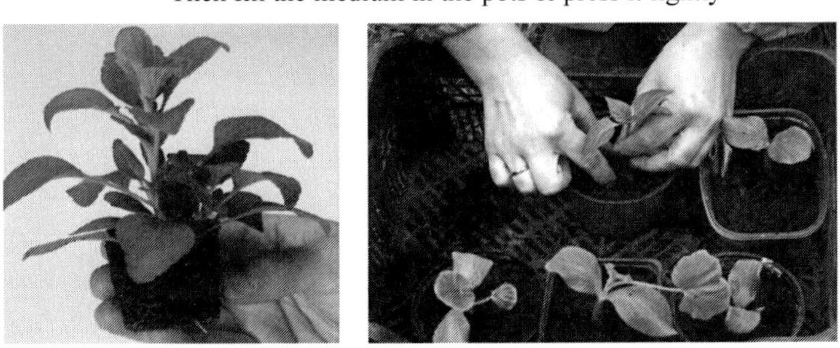

Place the plant with the ball of earth in the centre

Management of media and irrigation

Put pot mixture around the ball of earth, press as you fill up and level off, leaving one inch head space at top. Do not press over the ball of earth. It will break and damage the roots

Most of the vegetables are raised by sowing their seeds directly in containers. The seedlings of brinjal, chilli, tomato, capsicum, lettuce, cabbage, cauliflower, onion, parsley and leek are transplanted in containers. The seedlings can be raised in earthen pot or pans. A single healthy seedling may be transplanted in each container. More number of seedlings can be transplanted in each container for onion, lettuce and knol-khol, while two seeds of summer squash and 4-5 seeds of cluster bean, cowpea, okra and french bean can be sown in a single container. In case of radish, beet root, carrot and turnip, more number of seeds can be sown initially, then can be thinned out and only 4-5 healthy seedlings may be retained. A number of plants per pot can be raised in case of amaranthus, palak, spinach, fenugreek, mustard, chenopodium and coriander and the plants can be maintained for a long time and a number of cuttings (3-4) can also be taken.

Cultural practices

Plants in pots and containers need a lot of care and attention. It is essential to water the plants judiciously depending upon the season, kind of crop, size of the plant and size of the container. Plants need extra water during summer and the plants should preferably be irrigated twice a day. Too much watering will also lead to problems, hence we should strike an intelligent balance. The thumb rule for irrigation is that the top soil when scratched about one inch, if the lower soil remains damp, it indicates that no need for immediate irrigation. Due to evaporation, the top soil generally dries, even though the soil may have enough moisture to sustain the plant. Plants of cowpea, tomato, bitter gourd, lab lab, ribbed gourd need staking. Top dressing with nitrogenous fertilizers improves plant growth and yield of vegetables. This can be done by applying urea or DAP or ammonium sulphate in small quantities. In general, 5-10g of urea may be applied in moist soil once in a week or 10 days starting from 3 weeks after sowing or 2 weeks after transplanting. Heavy doses of fertilizer are very harmful. Immediately after fertilizer application, the plants should be watered. Hand hoeing and weeding helps in aeration in the root zone and help the plant grow healthy. Weeds should be removed gently with hands in leafy vegetables like amaranthus, fenugreek, spinach, coriander etc. Special operations like staking, thinning of excess plants, pruning of dried leaves and shoots, mulching may be followed whenever necessary.

Pest and disease management

Small sucking insects like aphids and jassids damage the plants in early stages of growth. Spraying of dimethoate @ 2ml/lit of water controls these insects. Fruit fly and fruit borer are serious pests in roof garden. They damage young fruits and make them unfit for consumption. The affected fruits should be plucked and destroyed. The plants can be sprayed with malathion @ 1-2 ml/litre of water. After spraying, the fruits should not be harvested for 7 days for consumption. Fungal diseases during rainy season may be controlled by drenching the soil with copper oxy chloride @ 2g/lit of water.

Harvesting

Vegetables are harvested at proper stage of maturity and used promptly. The crops grown in roof garden with adequate care are always superior in nutritional content, freshness, flavor and appearance. Poor quality and uneven ripening are caused by early harvesting and late harvesting result in extremely poor shelf life. Leafy vegetables should be picked up frequently when tender. Root vegetables should be pulled out while tender otherwise become pithy. Tomato is picked at colour break to ripe stage, brinjal and okra are picked when they are tender with optimum maturity. Rare vegetables like leek and fennel may not be available in market and hence can be advantageously raised in containers. Curry leaf dwarf papaya and gooseberry can also be grown in medium to big sized containers in a roof garden.

Limiting factors in roof garden vegetable cultivation

- Roof tops should be strong enough to hold the weight of the containers, plants and materials placed on them.
- Fertilizers specifically suitable for containers may not be available, i.e. fertilizers that contain micro nutrients.
- People may not be prepared to give daily care to a garden.
- It may be difficult to develop a uniform formula for making the gardens when only recycled materials are considered.
- Water may be a constraint.
- Good quality compost is usually not available unless people make their own.

Social impact of roof garden

Owing to the seat bound office jobs, there is hardly any physical activity in urban family life. Lack of exercise has led to a number of health hazards. Roof garden system of vegetable cultivation provides an alternative to people living in all kinds of dwellings like individual houses, flats, or apartments. A family can tend the garden as a team. This can be a healthy family time to interact and talk, while doing a useful activity from the point of health and economy. The physical exercise helps us to forget the tensions of office and our minds can freshen up.

With the passage of time, the traditional joint family system is breaking up in India and most of the families are fragmenting into nuclear families on account of struggle for money and job constrains. In families where grandparents are a part, the older generation feels left out. With a roof garden at home, even the old people can participate and feel one with the group and work rather than feeling left out. Thus, having a garden is not only a physical or monitory need but psychological too.

Other different kinds of vegetable garden on commercial scale are

- Market gardens
- Truck gardens
- Gardens for processing
- Vegetable forcing
- Gardens for seed production

1. Market gardening

This type of vegetable garden is intended to supply vegetables for the local market. Therefore, they are confined to the immediate vicinities of cities and towns. Such gardens are usually located within a radius of about 15 to 30 km from large towns and cities. Since these type of gardens are close to urban, lands are generally costly. Hence, intensive method of cultivation is followed. The gardener must be a versatile person as he is to grow a number of vegetables throughout the year to supply all kinds of vegetables to the market.

2. Truck garden

This refers to producing some special crops in relatively larger quantities for distant market. The location of this type of gardens is determined by the soil and climatic factors suitable for raising those particular crops. Cultivation is more extensive. The commodities are usually marketed through middle-men. In Tamil Nadu, growing of cabbage and cauliflower in Nilgiris and tomatoes in Periyakulam tract and sending them to distant places like Chennai are good examples of truck gardens.

3. Vegetable forcing

It refers to the production of vegetables out of season, in glass house. In England and other Western countries where winter is severe, vegetables especially tomato and cucumber can not be grown outside during cooler months. As they are required throughout the year they are grown under protected conditions. Special varieties have been developed exclusively suitable for vegetable forcing in crops like tomato and cucumber. In India, this type of garden has little scope and there is also no need for it since all vegetables can be grown throughout the year in one part or other of our country. Yet certain big companies have recently ventured in this field in order to capitalize the off season marketing of important vegetables like tomato, capsicum etc.

4. Growing of vegetables for processing

This is a type of garden which comes up around processing factories. They are intended to supply vegetables regularly to the factories. This type of garden is not yet developed in India due to lack of development of processing industries. Few factories which exist are solely depending upon the local market for their raw materials. In foreign countries, it is a specialized type of vegetable garden and they grow only those varieties suitable for processing only. The vegetables may be processed by canning, freezing, dehydration and pickling and fermentation. Higher and continuous supply and also the quality of the produce are very important in this type of garden.

5. Vegetable seed industry

Growing of vegetables to produce pure, true-to-type seed of good quality is another specialized activity. Soil, climate and disease free conditions are factors influencing location of seed producing areas. A thorough knowledge of the crop, its growth habit, mode of pollination, proper isolation distances are all of prime importance in seed industry. The handling of the seed crop, curing, thrashing, cleaning, grading, packing and storage are the specialized techniques and a thorough knowledge on these aspects must be possessed by the grower. In

India, this is an expanding industry having vast scope. At present, certified seeds are produced for sale in Government farms and authorized private nurseries. Seed law enforces standards, compulsory seed testing and certification in India. In and around Coimbatore and Madurai, some leading vegetable seed producing gardens have come into existence recently.

21

Micro Propagation and Genetic Engineering

Micro propagation in also called as *in vitro* propagation. *In vitro* is a Latin word, meaning 'in glass'. It refers to procedures done in controlled artificial environment, outside a living organism. These procedures naturally occur in the living system. Opposite to *in vitro* is *in vivo* which is also a Latin word, meaning 'in living'. Biological processes that occur within the living system.

Micro propagation refers to the multiplication of plants, in aseptic conditions in containers under controlled nutritional and environmental condition. Micropropagation is done in artificial growth medium from various plant parts like meristem tip, embryos, anthers, axillary buds, leaf section, shootlips etc. It is a method by which a true to type and disease free plant can be regenerated from a miniature piece of plant throughout the year. Some popularly grown crops from tissue culture are as follows -

Ornamental plants - orchids, lilium, gerbera, cordyline, begonia, ficus, etc.

Medicinal plants - *Catharanthus roseus, Coleus forskholli, Ginko biloba, Solanum trilobata*, etc.

Fruit Plants - banana, papaya, etc.

Vegetables - potato, aspargus, broccoli, etc.

Advantages

- Tissue culture helps in rapid multiplication of true to type plants throughout the year.
- A new plant can be regenerated from a miniature plant part, whereas, in conventional methods a shoot of considerable length is required.
- Large number of plants can be produced in culture tubes in small space with uniform growth and productivity instead of growing them in large areas in nursery.
- Plants raised by tissue culture are free from diseases.

- Tissue culture coupled with somatic hybridization (production of hybrid cells by fusion of two protoplasts with different genetic makeup) helps in evolving new cultivars in a short time.
- Micro propagation facilitates long distance transport of propagation materials and long term storage of clonal materials.
- Tissue culture methods are particularly effective in plants that do not breed true to type from seeds, or when the seeds are not viable (male sterile) or not available (banana) or in high value plants where propagation by conventional methods is not remunerative in large scale (orchids).
- Tissue culture plants have uniform growth in the field, which helps the grower to plan the intercultural operations, fertilizer application, harvesting and also transport of the produce to the market.
- Produce of assured quality is obtained.

Demerits of micro propagation

- The cost involved in setting up a laboratory is very high.
- The maintenance cost of a laboratory is very high.
- Tissue culture techniques require skilled manpower.
- Slight infection may damage the entire lot of plants.
- Some genetic modification (mutation) of the plant may develop with some varieties and culture systems, which may alter the quality of the produce.
- Field establishment percentage of tissue cultured plant is very low, making the plants expensive.

Methods of micro propagation: Different methods of micro-propagation are followed. These are named after the explant used. Callus culture, Cell culture, Meristem culture, Embryo culture, Protoplast culture, Shoot apex grafting, and Pollen grain culture. Few important ones are discussed below:

1. Callus culture

A piece of sterile plant tissue with living cells is transferred to a culture medium to induce callus proliferation. Sub culturing is then done; this means that few cells from the callus are transferred to a new medium. The callus cells can be inoculated again in callus medium to maintain callus, or in shoot inducing hormone enriched media to induce shoots. When shoots are induced, they are sub cultured to either shoot proliferating media or root inducing media. Once the shoots and roots are induced, a new plantlet is formed and it is incubated for growth and subsequently hardened and taken to main field for planting.

2. Cell culture

Cells are maintained in suspension cultures so as to produce free cells and are then subcultured to regenerate complete plant from a single cell. This technique is now useful to induce variability in plant cells and to select desirable cell variants and regenerate complete plants from these variants.

3. Meristem culture

This technique involves aseptic culture of shoot meristems on nutrient medium so as to produce complete plants. Most important application of meristem culture is the production of virus free plant which is genetically identical.

4. Embryo culture

It involves aseptic excision of the embryo and its transfer to suitable medium for development under optimum culture conditions. After the embryo has grown into a plantlet *in vitro,* it is transferred to sterile soil or vermiculite and grown to maturity in a green house. This technique is useful in the production of interspecific and intergeneric hybrids which cannot be otherwise accomplished and also in overcoming embryo abortion. Similar way ovule cultures are attempted in grapes. Ten days old fertilized ovules are excised and cultured *in vitro* to breed seedless grapes.

5. Protoplast culture

Protoplasts are the cells without any rigid cellulose wall, it has a plasma membrane only. Two protoplasts easily fuse and form a somatic hybrid. The hybrids are cultured in suitable media to regenerate cell wall and are again cultured in suitable medium for differentiation and morphogenesis.

Procedure for micro Propagation

Collection of explant

The small piece of plant used to begin a culture is referred to as an explant. The size, age and type of explant decide the success on *in vitro* propagation. Explants may be excised from shoot-tip, root-tip, axillary bud, embryo, endosperm, hypocotyls, callus and anther. The type of explant depends upon the method of tissue culture to be adopted. Explants are collected in containers having sterile water. The explant is taken to the preparation area.

Preparation of explant

Preparation of explant is done in preparation area. This area is also utilized for cleaning glassware, preparation and sterilization of media and storage of glassware. The explant is thoroughly cleaned and surface sterilized.

Surface sterilization

Explants collected from field grown plants harbor numerous fungi and bacteria, which when inoculated into a nutrient medium contaminate the entire *in vitro* system. Hence, sterilization is resorted to prior inoculation of explants. The efficacy of the sterilants used are found to vary depending upon the type of chemical concentration used, time of exposure etc. Commonly used chemicals for surface sterilization are –

- Sodium hypochlorite (0.5-2.0% *w/v),*
- Calcium hypochlorite (filtered 5-10% *w/v)* or
- Mercuric chloride (0.1% *w/v*).

A few drops of teepol are also added to facilitate better contact between the explants and the sterilant. Then the explants are rinsed 3-4 times with sterile distilled water.

Equipment needed in preparation area

Refrigerator, autoclave (120°C or 250°F with 15 PSI pressure), pH meter, gas or heating plate, stirrers and mixing device, filter, water purifier or distillation unit, vacuum pump (or) ultrasonic cleaner used to decontaminate explants and storage containers for flasks, bottles and petri dishes.

Auto-clave or pressure cooker

It is used to sterilize the media, containers, petri dishes and the various accessories required in the transfer operations. Normally sterilization is done for 15-20 minutes at 15 PSI pressure at 121-144°C.

Culture medium

Culture medium is used for inoculating the explant. Ingredients vary with kind of plant and propagation stage. It contains inorganic salts, organic compounds, vitamins, growth regulators, antioxidants (citric acid and ascorbic acid prevents the contamination), complex natural ingredients and agar (support agent).

Important media used for tissue culture

- Murashige & Skoog (MS) medium
- White's medium
- Gamborg G5 medium
- Linsmaier & Skoog medium
- Woody plant medium (woody plants)
- Anderson medium

Composition of some widely used media for plant tissue and cell culture

	White's medium	Murashige and Skoog's medium	Gamborg's medium
Macronutrients (mg/l)			
NH_4No_3	—	1650	—
KNO_3	80	1900	2500
$CaCl_2$ $2H_2O$	—	440	150
$MgSO_4$ $7H_2O$	720	370	250
KH_2PO_4	—	170	—
$(NH_4)_2$, $4H_2O$	—	—	134
NaH_2PO_4 $4H_2O$	16.5	—	150
$Ca(NO_3)_2$, $4H_2O$	300	—	—
Na_2SO_4	200	—	—
KCl	65	—	—
H_3BO_3	1.5	6.2	3.0
$MnSO_4$, $4H_2O$	7.0	22.3	—
$MnSO_4$, H_2O	—	—	10.0
$ZnSO_4$, $7H_2O$	3.0	8.6	2.0
$Na_2Mo\ O_4$, $2H_2O$	—	0.25	0.25
$CuSO_4$, $5H_2O$	—	0.025	0.025
$CoCl_2$, $6H_2O$	—	0.025	0.025
$FeSO_4$, $7H_2O$	2.5	27.8	0.025

Inoculation

The process of introducing explant, callus or shoot into growth medium is called inoculation. The inoculation room or chamber is used to transfer fresh explants into the culture medium, subsequent sub culturing and inoculating the shoot for rooting. This must be done in an aseptic condition. This is achieved by surface sterilization of the working table of the laminar air flow chamber with absolute alcohol followed by UV light for 30 minutes. The forceps and scalpel used for inoculation besides petri dishes are also first steam sterilized in an auto-clave at 121^{0}C for 15 minutes and later flame sterilized before each inoculation. The hands should be also cleaned with alcohol before use and the worker should wear sterilized mask.

Equipment needed in inoculation room

To transfer the explants into the culture medium under aseptic condition, laminar airflow chamber is a prime necessity. The other equipments used are UV germicidal lamp, alcohol lamp, forceps, dissecting needle, scalpel and blade, bottles, petri dishes and sterile water.

Laminar air flow chamber

This chamber is useful to perform all operations in aseptic conditions. Sterilization is achieved by the ultra-violet (UV) germicidal lamp fitted in it and by the flow of filtered air towards the person doing the operations, which prevents dust particles which carry microorganisms.

Sub-culturing

After inoculation, the explant proliferates and increases in volume. At this stage, it is divided into different components or parts and transferred into a fresh medium under above mentioned aseptic sterile condition. This process is known as sub-culturing. This sub-cultured mass should produce a shoot and root system which is dependent upon the type of growth regulator and its concentration used in the medium. It is generally observed that if the concentration of cytokinins is high in comparision to auxin in a medium, shoots are induced and on the other hand, when the concentration of cytokinins is low in comparison to auxin, roots are induced and at intermediate concentration the tissue grows as undifferentiated mass of cells, i.e., callus. This basic approach is widely used to regenerate a wide variety of dicotyledonous and monocotyledonous plants. The most commonly used auxins are 2,4-dichloro phenoxy acetic acid (2,4-D), 3, indole butyric acid (IBA), alpha naphthalene acetic acid (NAA) 3-indole acetic butyric acid and 2,4,5-trichoro phenoxy acetic acid (2,4,5-T). Similarly, the commonly used cytokinins are 6-benzyl adenine (BA), 6-furfuryl amino purine (Kinetin) etc.

Incubation

The inoculated cultures are incubated at 21^0-30^{0}C in an air conditioned culture room with 16-hour photoperiod (1000 lux) supplied by cool day fluorescent tubes.Specific environmental conditions are maintained in the incubation rooms as follows –

- Temperature – 21^0-30^{0}C
- Light 3 to 30W/m^2 PAR cool light or Gro-lux fluorescent lamps are provided
- Photo period 16/8 (day and night hrs)
- Humidity 30-60%.

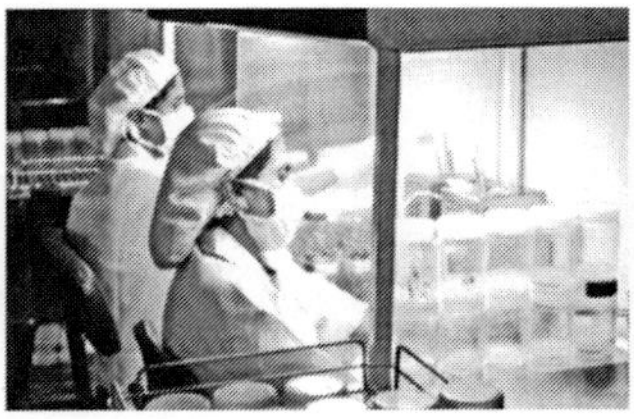

Equipment needed in growing area

The incubation room should have good air conditioner, fluorescent lamps and racks where the inoculated jar, tubes or bottles are incubated for further growth.

In the incubation room, the explant grows and proliferates. The callus grows, this may be sub cultured in inoculation room as callus culture to maintain callus lines or to shoot inducing media for organogenesis of shoots. In the incubation room, the shoots grow to the desired height and they are inoculated in root inducing media. The roots appear and the plantlets are taken for hardening before transplanting in open.

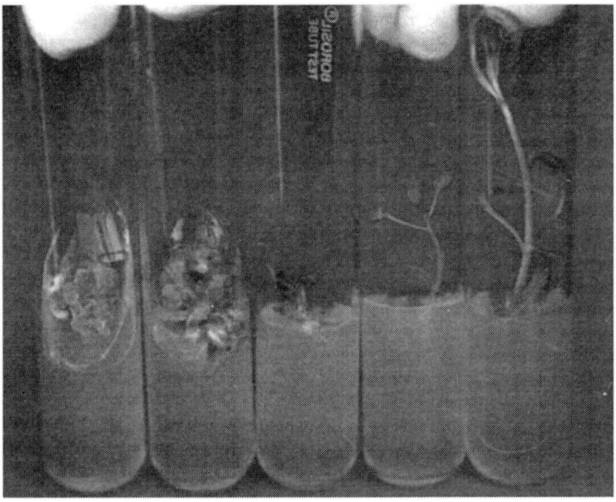

Acclimatization (Hardening) area

The plantlets developed in the culture tubes are acclimatized to a specific environment having a high humidity, a low light level and a constant temperature. Besides, the roots developed *in vitro* are hairless and delicate, requiring care during transfer from culture medium. To have a better survival rate, the plantlets may be transferred to container in mist chambers where relative humidity is maintained at higher level. Once new growth is seen, the plant may be slowly exposed to increased light intensity and temperature in stages, and finally planted in net pots/protrays in hardening chamber.

Collection of explants

↓

Surface sterilization of explants with sodium hypochlorite (0.5-2.0% *w/v),* calcium hypochlorite (filtered 5-10% *w/v)* or mercuric chloride (0.1% *w/v*)

↓

Inoculation (under laminar air flow chamber)

↓

Incubated at 21^0-30 ^{0}C+ 16 hrs light (1000 lux) in culture room + 30% humidity

↓

Sub-culturing (shoot multiplication in cytokinin rich medium)

↓

Rooting (root initiation and multiplication in auxin rich medium)

↓

Hardening (under mist chamber) and shade Net

↓

Ready for sale

Genetic engineering

Genetic engineering is a broad term referring to manipulation of nucleic acid of an organism. When the genes of the plants are artificially altered to achieve a desired effect, then the plant is called genetically engineered or genetically modified (GM).

The first step in making genetically modified plant is transfer of DNA into the plant cell. Foreign DNA can be incorporated in to a new plant, by following a few methods.

- **Heat shock**: The chimera plasmids are placed in a solution containing cold calcium chloride and normal host bacteria. On heating, suddenly to 42°C for 2-5 minutes the host bacterial membranes become permeable to plasmid chimeras, which pass into the cell.
- **Electroporation:** The host cells are subjected to a high voltage pulse which temporarily disrupts the membrane and allows the vector to enter the cell.
- **Viruses**: Since viruses have mechanism to infect susceptible cell and replicate themselves, a genetically engineered virus can deliver desired DNA sequence into the target host cell.
- **Gene gun**: Gold particles coated with foreign DNA segments are fired into the host cell.
- **Micro injection**: A cell is held in place with a pipette under a microscope and foreign DNA is injected directly into the nucleus using fine needle.

- **Liposome**: Vectors can be enclosed in a liposome, which are small membrane bound vesicles. The liposomes fuse with the cell membrane (or nuclear membrane) and deliver the DNA into the cytoplasm/nucleus.
- **Use of bacteria or virus**. The bacteria most commonly used are *Agrobacterium tumifaciens*. The relevant gene is transferred into the bacteria and the bacteria cells transfer this new DNA to the genome of the plant. This is called *Agrobacterium* mediated transfer.

Plant DNA can also be altered by mutational breeding and genome editing.

Isolation of the gene (DNA sequence)

The technique involved in recombinant DNA technology is to slice (cut) the desired DNA segment and introduce it into a vector (e.g., plasmid). This is achieved using a specific bacterial enzyme called restriction enzymes or restriction endonucleases. These enzymes function as endonuclease, which can cleave a DNA sequence at a specific site. These enzymes are named with three letters based on the species where it was isolated. For example, EcoRI is isolated from *E.coli (eschcrichia coli).*

Each restriction enzyme cleaves DNA strand at a specific site called recognition sequence or restriction site. For example, Eco RI recognizes the sequence GAATTC and cleaves it between G and A (G↓A).

Cloning

The sticky ends are generated by the same enzyme on vector as well as the target DNA are complementary and hence are cohesive.

The sticky ends of the cleaved DNA segment cohere with those of the vector, thus the cut DNA sequence can now be introduced into the plasmid. The cut ends are joined by DNA ligase enzyme and the introduced gene becomes a part of the plasmid. Ligase is an enzyme that covalently joins the sugarphosphate backbone of bases together. Ligase will join either "sticky" ends or "blunt" ends, but it is more efficient at closing sticky ends. The process of introducing foreign gene into a vector is called as cloning and the plasmid containing a cloned gene is called chimera. The DNA sequence that has been inserted into the vector is also called an "insert".

The chimera is then introduced into its host (e.g) *Agrobacterium tumifaciens* by various methods. Vectors carrying the genes must be incorporated into the living cells so that they can be expressed or replicated. The cells receiving the vector are called the host cell and once the vector is successfully incorporated into the host cell, the host cell is said to be "transformed".

Key advancements in plant research

Control of weeds - Genetic engineering has been successfully used to control weeds, plant diseases and insects. Developing the plants with herbicide tolerance confers selectivity and enhances crop safety, when herbicides are used. This will eventually reduce the use of herbicides and would be eco-friendly.

Insect Resistance - Insect resistance in plants has been mediated by *Bacillus thuringiensis (Bt)*, it produces a protein which is lethal to insects, particularly lepidopteran insects such as moth, butterfly and larvae. It disrupts the ion transport in the membranes of lepidoptera.

Disease Resistance - 'Coat – protein' mediated protection has been successful in case of Tobacco Mosaic Virus disease. Transgenic tomato with coat protein gene of Tomato Mosaic Virus (ToMV) shows resistance to viral infections. In papaya, a variety "Rainbow" was developed in Hawaii with resistance in Papaya Ring spot virus (PRSV) through coat-protein mediated transfer.

Regulated gene expression - Genes showing control on leaves, flowers, seeds etc have been identified, which help in targeting the gene expression accurately. Genes that respond to high temperature, frost, salinity, wounding, etc have been identified,

Gene tagging: Identification and isolation of new gene coding sequences have been explored in improving the plants

Gene mapping: Restricted Fragment Length Polymorphism (RFLP) genetic maps are available for many crops E.g) tomato. RFLP helps in improving the quality of a crop.

Constraints

- Approval from Government or regulatory body, food and environmental production agencies and other involved agencies is difficult to obtain
- Biosately issues about genetically modified crops have not been authentically dissolved. This issue needs experimental validation.
- Growing genetically modified crops in large scale may be a threat to natural biodiversity.
- Acquiring genetically modified seeds for large area seems to be a costly venture.
- A detailed study needs to be conducted to assess the stability of the performance of crops.

22

Ornamental Gardening and Landscaping

A garden has a number of elements which give shape to its breath-taking view. Both animate and inanimate elements are included to achieve a beautiful garden The living elements are lawns, rock garden, water garden, sunken garden, hedges, edges, shrubs, topiary, flower bed, specimen plants, pergola and arches, while the inanimate elements, also called as adornments, are stones, road, footpaths, seats, lamps, lanterns, steps, bridges, huts, swings, fountain, cascade, bird bath, clock tower, umbrellas, sculptures etc.

The most important garden elements are

i) Lawn

ii) Topiary

iii) Rock garden

iv) Water garden

i) Lawn

Lawn is a stretch of green grass-covered land, ideally 4 cm higher than the rest of the garden. A good lawn is the most eye-catching feature, it is the largest part of the garden and provides a beautiful background for the beds and other garden elements. A lawn has a smooth even surface or a velvety one, as per the preference of an individual and such a variation can be created by the choice of the grass and maintenance practices.

The lawn should ideally be planned in an area where it can get full sunshine, since most lawn grasses grow best in full sunshine nevertheless there are grasses which grow in the shade too.

The soil should be well – drained, devoid of weeds, stones and other inert material. It should have a gradual slant towards the drainage so that water stagnation is avoided in the lawn.

Land preparation for a lawn should preferably be done three months before planting. About, 40-60 cartloads of well decomposed Farm Yard Manure should be added per acre. The soil should be amended to get a loam texture.

The grass varieties for sunny location is Korean carpet grass, Mexican carpet grass, Doob etc., while for shady areas "Kentucky blue grass" or "Poa" is planted, which is a dark grass with broader blades.

The time of planting a lawn is when it is not too dry. Seed sowing and turfing are the two main methods of planting a lawn.

Seeds are sown evenly in shallow furrows in the prepared land. The seed rate is around 40 g/m^2. The seeds are small and should preferably be mixed with two parts of sand. The soil is kept moist till the shoots appear, which is about 30-40 days from sowing. Sprinkle irrigation should be done so that the seeds do not dislodge. If the lawn is planted by turfing, we should be very careful in choosing the turves. The turves should be weed, pest and disease free and should be uniform in colour. If a 30 cm square piece is lifted and shaken vigorously, it should not tear. The turf pieces should be placed in line starting from one corner of the lawn area compactly and the gaps should be filled with fine sand. The turves, after placing are firmly set by wooden tamper or base of the spade. After the turf placement, the lawn is swept with a stiff broom, to merge the edges and raise the grass.Turfing is the instant method of getting a lawn. Growing a lawn by dividing the turf is a quick and popular method. Square pieces of one-foot turf are neatly cut and spread to four feet lawn area. This gives a full lawn in about 3 months' time. This is cheap and best method of lawn plantation.

Regular mowing maintains the appearance of the lawn. Mowing is done weekly or fortnightly as per the grass growth. Hand weeding is simple and effective method to keep the weeds at bay. The grass encroaches the nearby beds, hence trimming the lawn edges is very important.

Evening hours are best to irrigate the lawns. In winters, irrigation should be done once a week or once in 10 days, while in summers, the frequency is increased to twice a week.

The lawn is fed twice a year, in June and October. Ammonium sulphate at the rate of 4-5 kg per 1000 square feet or urea at the rate of 2.5 kg per 1000 square feet is optimum.

Mowing for special effect

When a lawn is laid in a vast expanse, it looks monotonous if there is nothing attractive. To create a variation, it can be mowed in such a way that it gives a checked look of dark green and light green bands. The method of mowing is given in the following diagram.

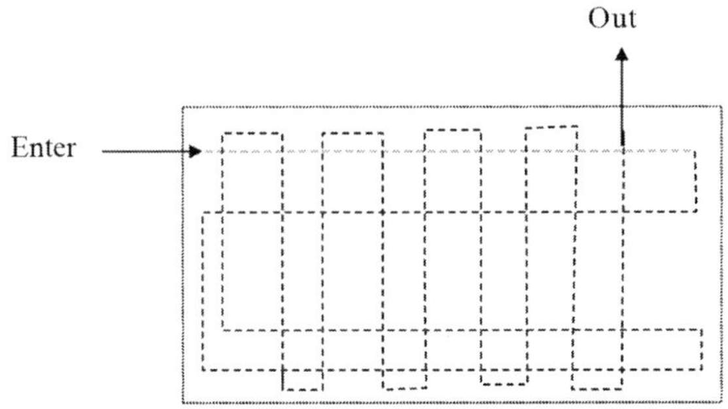

Pests are kept at bay by constant mowing and keeping the grass short. For white ants, 3-4" deep holes at the point of occurrence are dug and 25 ml of petrol per hole is poured and there after the holes are plugged with cow dung or earth mixture. Detrimental effects of black ants, grubs, larvae and cutworms are combated with BHC dust or insecticides with chlordane. Damping off is the major disease, which needs instant care. The seedlings topple and die, thus excessive moisture in the lawn is avoided.

ii) Topiary

The art of training compact plants to get a shape is known as topiary. Plants with compact foliage and which are hardy enough to withstand frequent clipping and shearing are suitable for topiary work. It takes many years to train and shape plants to desired shapes and sizes. Topiary is like a permanent feature of the garden, hence needs constant care.

It is possible to attain shapes of different types. Simple shapes such as a globe, sphere, dome and cube are not very difficult to achieve and can be made without the help of any pre-fabricated moulded wire model. Shapes such as bird, animal, etc., need preparation of a wire mould to train the shrub along the frame and clipping carefully over years to get the desired shape. A design with a broader base. For example, an animal standing over a dome or a cube is more impressive and manageable than standing on its own. It may take about 3 to 4 years to get a topiary of desired shape.

The shrubs most commonly used in India for topiary are *Clerodendron inerme*, *Duranta plumieri,* Bougainvillea, *Murraya exotica,* and species of *Hibiscus.* Generally, the difficult shapes of birds, animals, etc., are done with *Clerodendron inerme, Duranta plumeria,* and Bougainvillea. Thuja, *Cupressus macrocarpa, C. sempervirens, Putranjiva roxburghii* and *Polyalthia longifolia* are also used to shape domes, cones, spheres, umbrellas, etc. In temperate regions, Box (*Buxuss empervirens)* and English Yew (*Taxus baccata)* are widely used for topiary work.

iii) Rock Garden

In nature the plants on the rocks offer an eye-catching display, and in a garden, this is the replica, which is made to create the same effect.

A rock garden should be in a place where sunshine is available atleast for half the day, preferably in the forenoon. It should be away from any large tree and nearby hedges as their roots may interfere with the rock plants. A large rock garden will look out of place in a too formal garden. The site should be well drained and if natural drainage is not there, some artificial means should be provided.

The placement and selection of rock in a rock garden is an important factor. Rocks should be of local origin, porous and have a weathered look. Limestone from the quarry is a good choice they weather quickly. The stones should be of manageable size.

Before placing and setting the rocks the soil mound should be allowed to settle for some time and firmed by beating, if needed. The placement of stones will be such that it gives protection to the roots. The stones are placed in a slightly slanting position gently leaning backwards, not forwards, so that rainwater flows towards the plant roots. The rocks should slope in the same direction with the best faces visible, so that they represent natural stone strata. Pockets are created in between stones to place the plants. If the rockery is large, a broad path, about 45-60 cm, should wind through the rockery between the gorges and the valleys and make all parts of the rockery accessible. The path may be made of gravel or stepping-stones.

In a rockery, dense planting should be avoided. A pendant plant hanging down a projecting stone or a prostrate plant rambling along a large stone presents a pleasant view.

The best planting season is during the rains in the plains of India. In a climate as that of Bangalore planting is possible any time of the year except very hot weather.

The rock garden should be weeded periodically. It should be watered by sprinklers. The plants should be top dressed every year with a compost consisting of good sterilized garden soil, fine sand, and leaf-mould (or peat) in equal proportions.

People select the thorny cactus or some other hardy succulents for growing in the rocks. It is wrong idea to crowd the rockery with some cacti. Most of the delicate cacti cannot be grown in the open, especially in high rainfall areas and the hardy ones, which survive, do not bring any great beauty to the rock garden. Under tropical conditions, a selection of succulents, shrubs, ferns, and flowering annuals should be made for the rockery.

Some hardy species of cacti are *Opuntia, Cereus, Mammillaria* etc. Among the succulents, many species of *Agave, Aloe, Euphorbia, Furcraea, Gasteria, Kalanchoe, Sansevieria, Sedum, Sempervivum,* and *Yucca* can be planted in the rock garden. Some hardy ferns belonging to the genus *Drynaria, Nephrodium, Nephrolepis* and *Polypodium* can be grown in the shady portion of the tropical rock garden. Many shrubs are suitable for planting in the rock garden for example *Adeneum obesum, Asystasia coromandeliana, Azalea, Callistemon lanceolatus*, *Cassia alata, C. biflora, C. laevigata, Clerodendron macrosiphon, Crossandra, Cytisus, Daphne, Duranta plumieri variegate, Euonymus, Jatropha podagrica, Juniperus chinensis, J. horizontales, Lantana, Muehlenbeckia platyclados, Ruellia, Ruselia juncea, Thuja, Vaccinium* etc.

Some herbaceous plants like *Angelonia grandiflora, Aster, Crocus, Dracaena sanderiana, Fittonia argyroneura* (shade), *Impatiens holstii, I. sultanii, Oxalis rubra*, *Pedilanthus tithymaloides variegate, Pilea musoca*, *Portulaca,* miniature roses, *Sansevieria cylindrica, S. zeylanica, Setcreasea, Tradescantia, Zebrina pendula, Verbena, Vinca rosea*, *Vittadinia austrialis,* and *Zephyranthes* sp. can also be grown in a rock garden.

Although not considered as true rock plants, many of the dwarf annuals are quite suitable for the rockery. The following may be tried in groups at different pockets to bring colour to the rock garden. Dwarf *Alyssum, Antirrhinum, Bellisperennis, Brachycome, Browallia elata.* Candytuft, pantas, *Dianthus chinensis*, *Hymenatherum, Lantana, Mesembryanthemum, Phlox*, *Salvia*, *Verbena hybrida, Viola tricolorhortensis* and *Zinnia.*

iv) Water Garden

Water garden is one of the oldest features of gardening in India. In the olden days, it was customary to have lotus and lily pools in temples and stupas. Having a water garden has become rare in the modern landscape layouts because modern gardens are bound to small spaces, which does not allow the luxury of owning a water garden. More over a water garden needs constant care. Water gardens are various kinds, the informal and the formal.

Informal water garden is made in low-lying area of the garden, where there is a natural depression to hold water. The shape and size of such a pool will depend upon the general landscaping of the area and the slope of the land. The pond or lake will be used for planting true aquatics such as lotus, water lilies, etc. In the shallow waters near the banks bog plants are planted. On the banks at intervals, moisture-loving plants are planted to bring in a more natural effect. Most of the water plants do best in still waters. There are certain varieties of water lilies, which can be grown in water with gentle current. If the flow of water upturns the leaves or injures them, water lilies will not grow.

Unlike the informal pool, a formal pool is not just accommodated in a naturally low-lying area of the garden, rather it forms a main feature of the garden and occupies a prominent place and the garden is designed around it. Care should be taken that it is away from any large tree since the roots of the tree may cause cracks and crevices to the pool construction. The shape and size of a formal pool depends upon the size of the garden. The shape of the pool may be a simple rectangle or square, round or oval, kidney-shaped or any other indigenous artistic design. A formal pool is made of concrete, with a surrounding for growing the bog plants. The first step in constructing the pool is to mark out the design on the ground and then digging out the soil of the marked area to a depth of 30-90 cm in proportion to the size of the pool, plus the thickness of the concrete base. On completion of digging, all loose soil is removed and the base is made firm, as this has to support the tremendous weight of the concrete and water. Before any concrete work, the inlet and outlet pipes should be placed.

The usual depth of pools is 60 cm, but for growing lotus, a depth of 1 m is needed. In the plains of the tropics many species of water lilies thrive and flower even in 180-250 cm of water. But in places having severe winters, such depths will force the lilies to go dormant. Before filling up, the pool should also be tested against leakage.

Planting

The pool is filled with a mixture of loam, clay and Farm Yard Manure to the depth of 30-40 cm along with 50-75g of bone meal for every square meter area. The lilies or lotus are planted directly in the soil and water is filled gently to the crown of the plants. Water is added gradually as the leaf stalk grows. Alternatively, large pots can be filled with the planting mixture and slowly lowering in the pool of water. The water is gradually added as already explained.

In an already filled-up pool, where water level cannot be reduced, the planted pot is lowered to the crown, as the plant grows, it is slowly lowered down until the pot touches the base stably. Sometimes the plant dislodges from the pot and floats, in such a case the plant is kept in position with the help of stones, till it puts out root and anchors securely on its own.

Care of water garden

Water plants are hardy and grow well in full sunshine. In inadequate sunshine, the plants put forth vegetative growth, without producing flowers. 'Blanket weed', a green growth covers the surface and the sidewalls of the pool. This can be controlled by adding 30 g of potassium permanganate or 115 g of copper sulphate to each 100,000 or 125,000 litres of water.

Plants for water garden

Surface floating aquatics are *Aponogetondi stachyus, Euryale ferox, Monochloria hastifolia, Nelumbo lutea, Nelumbium uncifera, Pistia, Nymphaea, Orontium aquaticum, Azolla caroliniana, Elchhornia speciosa, Lemna gibba and L. minor, Trapanatans* and *Victoria regia.*

Oxygenators are *Callitrihe autumnalis, Creatophyllum, Elodea, Ranunculus aquatilis, Vallisneri aspiralis.* Marginal plants are *Acorus calamosvariegatus, Calla palustris, Cyperus alternifolius, Iris, Sagitarias agittifolia and Typha.*

Landscape Architecture

Landscape gardening is the technique of beautifying an area without much disturbing the land and soil. All gardens need meticulous planning so that all the elements of the garden are in harmony and merge together to appear as a whole system. A well-planned garden can be very relaxing and is a joy to the person enjoying it. Planning of the garden depends on the taste and choice of the person owning the garden, or the ultimate user. The purpose of the garden plays an important role in decision making. Some gardens need to be children friendly, while others should only be a quiet serene place for relaxing. Along with what needs to be done, planning also considers what can be done without making drastic changes to the naturally available landscape. Since laying a garden is not an everyday affair, long term planning and considerations are vital, such as prospective construction etc., so that tedious efforts in redesigning are avoided.

The area to be landscaped should be thoroughly studied. Observations should be made on existing flora, land topography, soil condition, fertility status, sun and shade during different times of the day. Presence of any tall building nearby should be noted down and also the shadow of the building in the garden area and its duration should be noted.

On the basis of the collected information, a rough sketch of the garden should be drawn, so that there is a preview of the design, before actually implementing it. All the construction work should be done in summers, while the planting work should be done during the rains. If the landscape of an old garden is to be modified, then it should be thoroughly studied and the design should be modified in such a way that the permanent features are not disturbed and changes are made in the temporary features like hedge, beds and may be lawn.

Garden styles

Every garden is laid in a certain style, which may be formal or informal.

Formal style

A formal garden is in strict geometric shape, all the elements in this look disciplined and well-manicured to perfection. A perfect symmetry is followed in the design so much so that the right side is the replica of the left. In a formal garden, there is no element of suspense, the whole garden can be seen from any point. The hedges and edges are well trimmed, following a definite pattern.

For designing a Moghul garden, a plan is put on the paper, the whole layout is in geometric patterns with the roads cut at 90 degrees. The flower beds are laid in the pattern of carpets. Fountains, pools or cascades are added. Moghul gardens are also formal gardens with splendid architecture and design. Moghul gardens are designed for the pleasure of the royal family or the tomb garden such as the Taj Mahal garden. The Moghul gardens follow the Charbagh design of paradise garden of Persia. Most of the Moghul gardens are divided into four parts by two channels of water, which form a pool in the centre. The four parts of the garden are designed by planting trees, shrubs and flowering plants.

Moghul gardens around tombs are very famous for example gardens around the Taj Mahal in Agra and around Akbar's tomb at Sikandra.

The Moghuls enjoyed terraces in the gardens so much so that they created artificial terraces where not available naturally. Eight terraces are preferred, to symbolize eight divisions of paradise, alternately the gardens may have seven terraces, signifying the seven planets. Moghul gardens have massive walls around them, with huge gates for protection from enemies and for enjoying the gardens in privacy.

Flowing water is almost an essential feature in all the Moghul gardens. Usually the water body is lined with blue colour tiles, to give an illusion of depth and clean. The canals have a pool, cascade or fountain at the terminal. A string of lights is added to enhance the beauty of the canal at night.

Baradari is the entertainment place, where the royals enjoy the dances. This area is covered from top and is integral component of a Moghul garden.

In Moghul architecture, each tree symbolizes different phases of life like youth, death etc, hence, trees are important in Moghul garden designs. Fruit yielding trees represent life and youth, while foliage trees like cypress symbolize eternity. The flowers in Moghul gardens are laid in geometric carpet like patterns. Scented flowers like rose and jasmine are preferred.

Informal style

The informal or the natural style of garden has its origin in Europe and is also called as British style of gardening. This design has an element of suspense, since there is no symmetry and obligation to repeat one side elements to match the other side, there is a room to display one's creativity and the visitor is forced to visit the whole garden. This garden is a large canvas of imagination of the architect and a number of elements can be added in different parts of the design, rendering each part a totally different ambience. A number of elements like seats, mushroom seats, umbrellas, koi pond, cascades, rockery, water pool, fernery, vertical garden, rose garden etc. find place in this setting without any restrictions. This garden is purely a product of expression of personal imagination and preferences.

Japanese style

Japanese style of gardening has very a few choices of elements. Stones of various shapes, sizes and colour, stone artfacts like bird baths, lanterns and statues are used. Sand and bamboo are also liberally used in the Japanese style of gardening. A tea house in the garden is like a relaxation centre. Too many colourful flowers and leaves are avoided to attain focus and peace. Water element is included where ever possible. This garden is laid totally in informal style.

The important forms of Japanese gardens are Hill garden where hills and water are the main components, Flat garden which lacks ups and downs. Tea garden, which has a teapoy, low tables, mats, wooden walls etc. for a tea party, Passage garden which focuses on decoration of passages and Sand gardens, where sand is used to create attractive patterns, just as in Zen garden.

In Japanese garden water is included in many ways, for example the ponds are in irregular shape, the streams are decorated with large stones, waterfalls are partially screened with bamboos and the fountains are laced with bamboo planting. The wells in a Japanese garden are of a geometric pattern and are very ornamental.

A Japanese garden focuses on islands. They are made in different shapes, for example the "Elysian Isle" (Horai-jima), which is devoid of a bridge and is constructed in the middle of a lake in the shape of a tortoise. The "Wind-swept Isle" (Fukiage-jima), is also an ocean isle and constructed in a similar way. Sea rocks adorn the beach.

Bridges in a Japanese garden are made with stones, wood etc. The bridge is used for a leisurely walk to admire the garden. To force a visitor to spend more time in the bridge, a bridge is made as long as possible with curves.

Too much colourful vegetation does not find a place in a Japanese garden. Flowering plants like *Magnolia grandiflora,* chrysanthemums, carnation, irises, lotuses, peonies, orchid and creepers like *Clematis, Ipomoea hederacea* and

Wisteria sinensis are sparingly used. Green colour is brought in by the Bamboos and some needle leaved plants.

Types of gardens based on utility

Gardens are made to suit personal requirements. Some gardens are laid for children to play, some for the family to relax and some to hold grand functions. Some broad kinds of gardens are:

Family garden

This garden is made with a sitting area for the family and a play area for the children. Thorny plants are avoided, artefacts with sharp edges are avoided but flower bed, lawns, flower borders, hedge etc. are included.

Small outdoor garden

This is a garden where small area is available and design is created by using plants in pots in flat or vertical gardens.

Screened and secluded sanctuary

This type of garden is made with the view of privacy. Here tall hedges, screens, pergolas are used to conceal the view of the garden.

Planning a garden

Laying out a garden needs meticulous planning. Many aspects have to be considered. After selecting a site, it is important to gather all the information with respect to the altitude, frost pattern, monsoon months, rainfall patterns, temperature fluctuations and sunshine hours, in the past one year. Information about soil kind, water pH, water quality and availability is required. Assessment of the nearby building structures and permanent trees should also be done, to know about the shadow they cast on the site and for how many hours. Any high and low-lying area of the site should be noted. Availability of water and labour should also be considered and plants may be chosen accordingly.

A design which does not entail drastic modification of the site should be chosen. Naturally high area can be converted to mounds or plants that do not require too much moisture can be planted there while the low-lying area can be used to make a water body, or for bog and marsh planting. Every element that has to be included should be imagined in combination with the other elements to create harmonious blending. It is always good to know the preference of climatic conditions of plants before combining them. Before various plants are put together, we should also know their natural habitat so that we do not put two contrasting habitat plants together, for example the cactus and maiden hair fern.

When designing a garden, the preferences of the owner are very important. After jotting down the owner's preferences, a landscape designer should ascertain the feasibility of the client's choices, give alternatives where ever required and make an assessment of the maintenance constraints.

After all the information is collected and all the elements are decided, a rough plan is made to see how all the elements will look when put together. After the rough, a detailed to-scale plan should be made, which gives the exact idea how the plan will look, with every element in proportion.

The garden execution should be done in a logical manner. Designing along the topography with the least changes is always better. Before planting, all the construction work for all 'hard' elements should be done during sunny days and all the planting should be scheduled for the rainy season.

A beautiful garden rejuvenates the soul. A garden should look beautiful always and this is possible when it has been laid out intelligently. Most of the gardens look good when freshly laid, but loose the beauty with time, this is because the garden has flora which needs a lot of maintenance and labour is scarce in the area, or during laying out, seasonal plants have been used in a great number, which wither away with time and the replanting becomes cumbersome. Any professional should let the client know what they should expect in their garden after it has been done. The client should be given a list of scheduled operations for maintenance.

23

Flower Arrangements, Floral Arts and Bonsai

Flowers of the garden are brought indoors to enjoy the beauty. Fresh flowers are arranged in various styles according to one's choice. There are two major and distinct styles of arranging fresh flowers, namely the European and the Japanese. Both the styles are greatly influenced by the place where they have been originated. Flowers are also dried and used for different arrangements. Fresh flowers are used as ornaments and also for making floral patterns on the floor called Rangoli.

Western flower Arrangement

A mass of flowers and foliage arearranged in different patterns keeping the following principles in mind.

- To draw attention, one focal point is created in a small arrangement, while 2-3 focal points are created in a large one. It should be in the lower one third of the arrangement.
- The arrangement should look balanced and proportional. The arrangement should be bulky at the base and taper delicately as it progresses.
- The colour should be the darkest or lightest at the focal point, from where it should be lighter to lightest or dark to darkest towards the periphery.
- The foliage and flowers should look pleasing together and the whole arrangement should stand out as a single piece.

To create a floral arrangement, a container, leaves fillers and flowers are required.

The leaves, fillers and flowers are arranged to get a form. The shape of the material, the texture of the material and the colour matters a lot. The colour combination should be pleasing to the eyes. Matching colour arrangement or an arrangement with contrast and bold colours give rhythm. The basic categorization of colour is as follows -

Primary colours: yellow, red and blue

Secondary colours: orange, green, and violet

Tertiary colours: red, orange, blue, green, etc.

Selection of material

- Container - It is very important element for the design. It should hold enough water and large enough to bear the weight of the arrangement.

Selection of flowers

Flowers of different shapes and sizes, arranged uniquely, are seen in a plant. The flowers which are as spike like the gladiolus, tuberose etc. are called the line flowers. The flowers which are single and in different shapes like anthurium, spathyphyllum, rose, bird of paradise etc. are called form flowers. Flowers which appear in a bunch like the floribunda rose (flowers in a bunch) are called the mass flowers and the flowers which are very small in size, but are in bunches are called the filler flowers like solidago, gypsophilla, statice, limonium etc.

Selection of foliage

The foliage which is single and bold, which makes the base of the arrangement is called mass foliage. E.g., Cycas, palm. The foliage which is thin but more in volume is called filler foliage e.g., Ferns and asparagus.

Other things required for making a flower arrangement

- Floral foam
- Scissors
- Long spout watering can

Types of floral Arrangements

Triangular arrangement

Triangular arrangements are of two types, equilateral triangle design and asymmetrical triangular design. For an equilateral triangle arrangement, the height and width is same. A tall flower is placed in the centre and two flowers are place on sideways which are together as wide as the tall flower. A short stemmed flower is placed as a focal point, just above the lip of the vase. This basic structure is complemented with foliage and filler flowers to get the final result. Asymmetrical triangle arrangement on the contrary is worked up as just a triangle, with undefined height and base. This arrangement suits a corner table.

Circular

This arrangement has no focal point. A number of matching or contrasting colour flowers are used in a circular manner and then foliage and fillers are used to create balance and appeal. The vase for this arrangement should be circular. Since this arrangement looks same all directions, it suits a centre table where it can be viewed from all sides.

Crescent designs

The crescent design is constructed in a half-moon shape. It is a centre table arrangement. The crescent height on the left is two third portion of the entire crescent. A curved foliage is placed to the side (usually left of the centre) so that its tip nearly reaches the visual centre line of the design. The flowers in this design are used in a way that the small flowers make the tips of the crescent while larger flowers make the focal point.

Hogarthian curve

It is an asymmetrical design in form of an 'S'. Long-stemmed vases are used for this design, because a portion of the floral line extends below the rim of the container. The upper curve of the 'S' consists of two-thirds the height of the total design. The lower portion of the curve is constructed with either dainty, dangling flowers or cluster of grapes.

Radiating designs

Radiating designs are shaped like a fan. Line flowers or foliage are used to form the outline. The height of the arrangement is established first then the width, to construct a fan.

Horizontal designs

The horizontal design is an inverted crescent. It is meant to be for a centre table. In this type of arrangement, the horizontal length is 1 ½ -2 times the length of the container. The skeleton flowers are placed midway between the back and front of the container. A focal point is established on each side to attract attention to the design. Fit to be used with candles for candle light dinners.Tree candles, with the tallest in the centre are placed for appeal. All foliage and flowers located near the candles should be low enough so they will not be burned as the candle is shortened by the flame.

Ikebana

Ikebana or Japanese arrangement is very versatile and uses a wide array of materials such as moss, root, leaf, lichen-covered branch, fruit, fungus and flowers. The Japanese rule of thumb is the fewer flowers, the better. A fully open flower, a half open bud and the tight bud may be used to symbolize past, present and future. The flowers and branch tips all point upwards, towards heaven.

Materials for Ikebana

i) Containers

A container is an integral part of any arrangement. It should always be in harmony with the arrangement. Metal containers are best for winter arrangements, while clay pots bring out the beauty of spring and autumn flowers. Summer flowers are ideally suited to baskets.

ii) Bases

A base, or *dai*is placed beneath a vase or bowl. For a square or rectangular container, a circular or irregular base is apt while for a round container, a square or rectangular base suits the best.

iii) Tools

An ideal Japanese tool sets consisting of a pair of scissors, a saw, a very small chopper, a two-edged knife and a syringe.

iv) The kenzan

The *Kenzan* or the pinholder is important in securing the arrangement material. Basically, a *Kenzan* must be capable of bearing the weight of branches.

Preparing Ikebana

Cutting the plant material

- The plant material to be used in Ikebana should always be cut diagonal.
- Flowers are cut straight across the stem.

Trimming the plant material

The plant material is trimmed, keeping the main line of arrangement intact. Anything that is outside its urview or which is diseased, disfigured and drooping, should be cut away.

Positioning the material

There are three main placements in ikebana and they are shin (heaven), soe (man) and hikae (earth) *shin*is the longest and most important line, *soe* is the medium line, and *hikae* is the shortest line. Shin, soe and hikae are placed at the angles of 10°, 45° and 75° with reference to an imaginary line rising vertically. Other material is added only after these three are in position. Additional branches and flowers are called *jushi,* which supplement the above three placements. *Jushi* are always shorter than the main stems.

An ideal arrangement would have the tips of the main stems pointing to one another so that the eye of the viewer is drawn to the centre or heart of the arrangement, where the *jushi* is positioned. The arrangements are viewed from the front thus should have a three-dimensional look.

In standard arrangement, the height of *Shin* is equivalent to the width of the container plus the depth and 50% of the above value. *Soe* is three-quarters of *shin* and *hikae* is three-quarters of soe.

In large arrangements *shin* is equivalent to twice the sum of the width of the container plus the depth, *soe* is three-quarters of shin and *hikae* is half of soe,

In small arrangements, *shin* is equivalent to the width of the container plus the depth of the container, *soe* is three-quaraters of *shin* and *hikae* is three-quaraters of soe.

All of the foregoing measurements are applicable to both *moribana* and *nageire* arrangements.

Basic styles of Ikebaba

i) Basic moribana

The 'moribana' arrangement is low and done in shallow containers. Basic *Moribana* arrangements should be placed on a low table against a simple plain background. It can be made with any material but should look most natural.

ii) Basic upright moribana

In this the pin-holder is placed at left front corner of the container. The *Shin* is fixed at the rear end of the pin-holder at an angle of 10° to 15°, with a forward tilt of 10°. Soe is arranged on the left of the *Shin* at 45° with a forward angle of 50°. The *Hikae* is positioned on the right side at the angle of 75° with a forward inclination of 30°. The Jushi are now fixed and it should be the shortest.

iii) Slant moribana

In this arrangement the pin-holder is positioned at the rear of the container towards your right end. Here, all the angles and inclinations remain the same as above, except that the angle and inclination of *Shin* is interchanged with that of *Soe* and vice versa. *Hikae* remains in the same position as above. The Jushi are arranged as per rules. As the tallest main line Shin is placed at a relatively slanting angle of 45°, the total effect also becomes slanting.

iv) Basic Nageire

Nageire arrangements are done in tall, upright containers. It is a very natural looking style.

v) Upright Nageire

The opening of the vase is divided into four quadrants and only one quadrant is used at a time for arranging the flowers. All the angles and inclinations for the three main lines are exactly similar to that of upright *Moribana*. The Jushi are arranged in the same quadrant.

vi) Morimono

The arrangement in combination with fruit and or vegetables and flowers are termed by the Japanese as *Morimono*. These are arranged on a wooden tray, basket, or a shallow bowl. Fruit and vegetables such as grape, orange, grapefruit, pineapple, apple,red tomatoes, capsicums, brinjals and carrots are suitable for this purpose. This also follows the three main lines viz; *Shin, Soe* and *Hikae*.

vii) Jiyu-bana (Free flowers)

It is arranged in both *Moribana* and *Nageire* styles in combinations of wood, metal or any other material. In this style, though the prescribed rules about angles and measurements are ignored to a great extent, the basic themes "Heaven", "Man", and "Earth" are followed.

viii) Zen ei-ka (Abstract style)

In this arrangement, plant materials are used in combination with inert material. In this style, generally straight materials of uneven heights are used.

ix) *Zen ei-bana* (Avant-garde)

This style looks like a piece of art where a beautiful sculpture is created using wood, stone, rocks or metal to depict a natural scenery.

Dry flower arrangements

Fresh flowers are expensive and are not available all the year round, so in the season, when there is glut, the flowers are dried and arranged whenever required. It can be used again and again, creating a new form every time it is used. Dry flowers and leaves retain the natural colours, shape and size. Material for dry arrangements include cultivated flowers, vegetables, wild flowers, seed pods, leaves and grasses. Some unusual flowers, leaves, pods etc. also make good material, if dried properly.

Dried flowers for colour

A well dried material retains its colour and dries uniformly with out the colour being faded. These are some of the plant materials that can be dried for their colour quality.

Red	-	Cockscomb, Peony, Pomegranate, Zinnia.
Pink	-	Gladiolus, Larkspur, Peony, Snapdragon, Statice.
Yellow	-	Acacia, Goldenrod, Marigold, Strawflower, Yarrow, Zinnia.
Blue	-	Cornflower, Delphinium, Globe Thistle, Hydrangea, Larkspur.
Green	-	Foliage, Grasses, Seed Pods.
Orange	-	Marigold, Strawflower, Zinnia.
Violet	-	Gladiolus, Heather, Lilac, Statice, Stock.
Gray	-	Artemisia, Bayberry, Dusty Miller
Brown	-	Canna, Cones, Dock, Seed Pods, Traveler's Tree
Tan	-	Grasses, Leaves, Seed Pods, Wood Roses
Black	-	Baptisia Pods, Magnolia Leaves, Teasel
White	-	Everlasting, Honesty, Peony, Statice

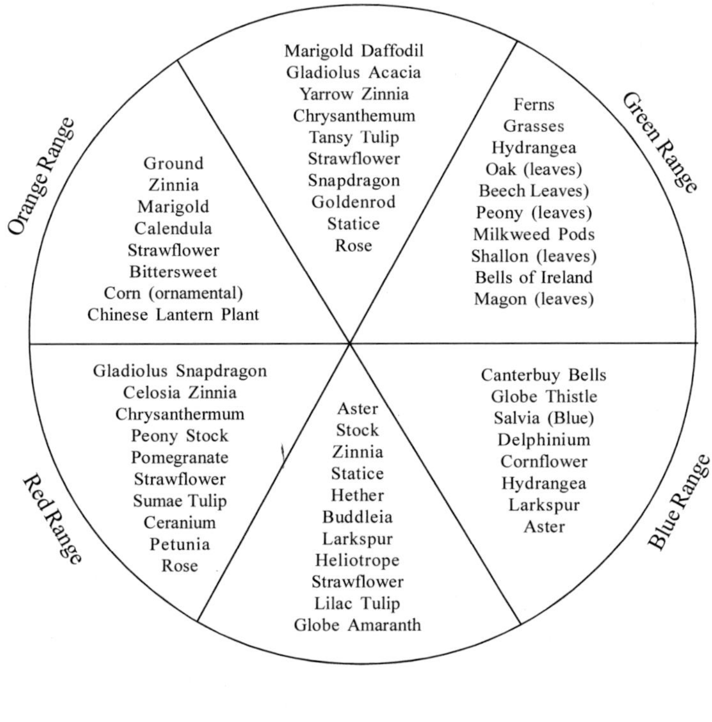

Collection

1. Watch timing. Unless otherwise specified, gather materials just as they mature.
2. Collect during a sunny day.
3. Strip foliage from all flowers to be dried.
4. The material should be kept in well ventilated shady area.
5. To dry in a shape, set the material in shape gently while the material is still fresh.

Methods of Drying

i) Upside-down method

Bunch the material together and hang upside down to dry. Flowers with dense petals or big head should be dried individually e.g; marigolds, roses, cineraria, sweet william and hydrangeas.

ii) Burying in sand or borax

For this method, dry beach sand or borax is used, though sand is the best. Damp sand or wet material will result in formation of spots. Layer a box with paper, to a thickness of an inch before putting 3-4 inches of dry sand on it. Place the heads of flowers and plant material in it and cover with sand or borax, in such a way that the material has space for air circulation.

It takes a day or two for the material to dry. A sample can be drawn to check. The material, if kept longer, develops spots and brown colour. The sand or borax can be used a multiple time.

iii) Glycerine method

Glycerine method is used only for thick foliage and flowers with thick petals e.g., celosia. The material is first washed to remove dust and dirt. The defective material is pruned and the lower part of the stem is crushed with a hammer to split the bark and the fiber. Now in a well-mixed solution containing one part of glycerine and two parts of water, place the stems of flowers and allow to stand for about two weeks. Circulation of air is necessary while material is drying. This prevents leaves from drying out, before they have absorbed enough solution. It is also wise to wipe leaves occasionally with a damp cloth, or with glycerine, as this helps to check drying before the solution has reached the edges. The glycerine solution can be kept away for subsequent use.

iv) Pressing

The technique of pressing the material flat, dries the material and the flowers do not retain the natural 3D shape. A good way to press a number of leaves or flowers is to place them between several layers of newspapers. A heavy weight is placed on the newspaper. The material is dried in three weeks.

v) Skeletonizing

The material is boiled with a pinch of soda in water for about thirty minutes. When cooled down, it is spread on newspaper. With a blunt knife, the material is scraped off gently to expose the skeleton. The material is left to soak in bleaching solution for an hour. After rinsing with clean water and dabbing with a dry cloth, the material is kept in sheets of newspaper for a day. The skeletonized leaves hold shape well and are sturdy, they appear delicate, misty and white.

Storing the material

The dried material should be stored in cool and dry place.

Types of design

The designs are in 'line' or 'mass'. In a 'line'arrangement, the use of material is scanty. There is a definite outline but the ends fade out and the focal points are bold. Specific material emphasising the line form are used.

In a mass arrangement a lot of material is used. Tiny spray flowers are used in bunches to bring in balance. The heavy material is placed in the centre, with gradual decrease in size towards the edges.

Dry flowers are also arranged in the same styles as the fresh flowers, e.g. vertical, oval, triangle etc and just as the fresh flowers, dry flower arrangements also need to have a balance, harmony, focal point, rhythm and unity.

How to locate the focal Point

The eye is drawn to the focal point by lines established in the design. Lines should be placed close together and appear to radiate from this spot. It is the boldest point which holds the interest of the viewer.

Colour Harmonies

The most commonly used colour harmonies are -

1. Monochromatic –It has a single hue in light and dark tones.
2. Analogous-blending-Closely related colors, those adjacent on the color wheel.

3. Complementary-contrasted hues - These are opposite on color wheel, as green leaves and red berries.

Direct: colors opposite on color wheel, as green-yellow with red-violet.

Near: one color plus another close to its direct complement, as green with violet-red.

Paired: two sets of direct complements, as red and green with yellow and violet.

Split: one color and the two colors adjoining its direct complement, as orange with green-blue and violet-blue.

Floral Art

The arrangement of the flowers which create a pleasing effect is known as the floral art. Floral art is an age-old practice around the globe and it differs as per the occasion, availability of material and taste of an individual.

Floral designs

A) Floral garland

It is one of the oldest methods of using flowers. Floral garlands are worn around the necks on special occasions like festivals, marriage or felicitation, they are also put around the statues of great individuals and deities. Garlands are also used for decorating a stage or a temple. Usually fragrant flowers are preferred for garlands like rose, jasmine, tuberose etc. Basil leaves are also used intermittently to create colour and also because it is considered medicinal and auspicious.

Besides garlands, the other important floral ornaments are bangles and floral crowns, which are worn by dancers or some royal stage characters.

B) Hair decoration

Flowers woven in different styles or singly have been used as hair adornment for centuries. Usually flowers with fragrance such as jasmine, tuberose or champa are preferred. In India, use of flowers on hair by women is noticed in Southern parts, including Maharashtra. Women wear flowers singly on the hair, as a rose or champa, but the most exquisite way to wear is in the form of veni, or a formation where the flowers are woven together to be worn along the length of the braid. It is also worn around a hair bun and in such a condition, it is known as gajara.

C) Rangoli

In India, rangoli is made with flowers or dry colours on special occasions, in front of houses, temples or convention halls. When floral rangoli is made, the flowers and flower petals are arranged in spectacular patterns of vibrant colours. The 'Poo Kollam' or floral rangoli, is famous in Kerala during onam.

D) Bouquets

A bouquet is made when flowers, foliage, filler flowers and filler foliage are arranged and tied together to make a handy arrangement. The construction of a bouquet is finished by covering the bouquet and tying it with an artistic golden, silver or satin bow. Bouquets may be of many types and shapes. There are two ways of making a bouquet, with frame and without frame. Bouquet is given for felicitation, for marriages and birthday etc.

E) Buttonholes

A buttonhole is a small assembly of flowers and foliage, usually worn by males in their coat collars. A portion of an artistic leaf like asparagus/thuja is taken and tied with a small rose bud, or single orchid, daisy or chrysanthemum. The base of a button hole is tied with aluminium foil to make it sturdy so that insertion in the button hole of the coat is easy.

F) Corsages and specialty pieces

A corsage is a specialized arrangement of flowers which is to be worn. It is more of a floral adornment. Since it has to be worn in different parts of the body, it is made flexible by using tape covered wires. Flowers and some foliage, along with satin ribbons, beads etc. are used to construct a corsage. A well-designed corsage should compliment the dress and complexion of the person who wears it.

Styles for Florist's Corsages

i) Shoulder corsages

A shoulder corsage is crescent shaped and is worn on the left side or on the right or around the front collar. The largest flowers are placed in the centre of the crescent and then the size tapers downwards consequently the smallest buds are at the tips of the crescent.

ii) Wrist corsages

A wrist corsage is worn as an accessory for dances. To make a wrist corsage, real flowers with artificial leaves and pearls or pollens are used. All cut wires should be covered neatly with floral tape.

iii) Corsages for the hair

Some women prefer to wear flowers or corsages in their hair. Raised flowers are not preferred to make a hair corsage. Flat flowers, bows, nets along with floral tape are used. It is woven with a hair pin or a small comb, so that it can be worn conveniently.

iv) Nosegay arrangements

The nosegay is meant to be carried in hand rather than worn. The flowers are arranged in a sphere along with a handle, which makes it convenient to carry.

G) Funeral designs

Fully opened, fresh looking, bright and beautiful blooms are used to make funeral pieces. Different kinds of funeral designs are enumerated below:

i) Basic casket spray

ii) Wreaths

Bonsai

Bonsai making is a wonderful Japanese art. In Japanese "Bon-Sai" means shallow pot. The experience of making a bonsai is thrilling since it is challenging to grow a miniature tree in a shallow pot.

Selecting a plant for bonsai

For making a bonsai, one should select a plant with small foliage and woody trunk, which looks like a grown tree, in its miniature form.

The list of ideal plants that can be chosen from is as follows.

a) Conifers and thin leaved trees-Mayflower (*Delonix regia)*, deodar (*Cedrus deodara*), fir (*Abiespindru)* etc.

b) Broad leaved tree species-Peepul (*Ficus religiosa),* banyan tree (*F. bengalensis*), ginko (*Ginko biloba) etc.*

c) Flowering and fruiting trees- Magnolia, citrus / orange, plum, pear, almond etc.

d) Shrubs-Bougainvillea, hibiscus, duranta, lagerstroemia, pomegranate etc.

e) Creepers-*Clerodendron splendens*, wisteria etc.

f) Foliage plants-Bamboo, jade plant (*Crassula argentea*) etc.

Season of planting

Monsoon is the best time for lifting a sapling and planting it in a pot to train it as a bonsai or seeds can be sown in an appropriate season to initiate a bonsai.

Manuring

Utmost care must be taken in the use of fertilizer. A handful of Farm Yard Manure may be used once a year. For plants which shed their leaves one spoon of the commercial formation of 0:10:10 NPK may be used during autumn, sparingly.

Watering

The best way to water a bonsai is by placing the pot in a large container. Watering from top will disturb the soil and the root system, especially in a fully made bonsai where the soil depth is very less.

Shaping the bonsai

The dead and extra branches, which do not fit the norms of the chosen style are removed. A slant cut is given for shaping, because a blunt cut gives unfinished and crude look to the bonsai.

A soft copper wire is used to train the branches in the required shape and form. The wire is worked to the length of the branch in a spiral manner and then the wire clad branch is bent to achieve the desired shape. Tying of the copper wire should neither be too tight or too loose. To get a cascade look, small weights can be tied on to the tips of the branches and the weights may be gradually increased.

Pruning

In bonsai, the pruning of root is as important as pruning the shoots. The tap root should be gradually cut to reduce the depth of the pot because a good bonsai is in a shallow pot or plate. The roots should be cut during the monsoon, when the pot may also be reduced. The shoots also should be trimmed once a year but should be trimmed after flowering or fruiting. The canopy should be twice or thrice the size of the pot. The cut should be smooth with a sharp tool.

Bonsai Styles

1. Formal upright style

The formal upright style is the basis of the main bonsai styles. The tree has a straight tapering trunk. The trunk of the tree is divided into three equal parts. The top one third of the trunk has three or more small branches which turn upwards and provide the dense foliage to the upper part of the tree. The second one third of the trunk carries the three main branches, the first from top is behind the trunk while the two lower ones are on either side. The lowest one third is devoid of branches.

An oval or rectangular pot suits this kind of bonsai and the plant is placed in the centre of the first half of the pot, keeping two thirds of the pot free for spread of the branches. If a round pot is used, then the plant is placed in the centre.

2. Informal upright style

This style is just like the above style, except that the trunk is slightly curved or slant.

3. Slanting style

In this style, the tree is slanted to one side of the pot and the lowest branch spreads out in the opposite direction of the slant. Pot placements are just as in the case of upright style.

4. Winds wept style

This style creates a look of a tree that has been swept by the wind.

5. Semi and full cascade styles

In semi-cascade style, the tree trunk grows upwards and then bends to one side and downwards to a position not below the base of the pot, while in the full cascade style, the trunk is allowed to grow upwards for a short distance and then allowed to fall steeply over the side of the container till it may be well below its base and therefore must be placed at the edge of a ledge or table.

6. Broom style

This is an upright style with all the branches appearing from the upper part of the trunk and spreading outwards like a fan.

7. Bunjin or literati style

This style of bonsai is relatively new, there are no rules but the trunk is kept slender and curved or slant.

8. Multiple trunk style

This is a more natural looking style, where the tree may have two or more trunks which may be straight, slant or in combination.

9. Group plantings / forest style

Two or more trees are planted together to give this kind of look.

10. Rock plantings

The plant is placed in the hollow of a stone or stones are placed in the roots, giving the look of stone clasped roots. A plate like flat container is used.

11. Mame' or miniature bonsai

These are small bonsai which are 8 to 20 centimetres tall.

24

Horticultural Therapy

Horticulture is an art and science of growing plants. It is the practice of engaging ailing people in various horticultural activities such as gardening and flower arrangements in order to improve their physical, mental and spiritual health. Research confirms that health benefits escalate manifold when people connect with nature and plants by viewing and/or interacting with them. Major benefits of horticultural therapy includes sensory simulation and mood enhancement, mental relaxation, improvement of memory and concentration and an overall feeling of wellness. Horticulture therapy also helps in cultivating positive emotions and provides a meditative mood. It eases both physical and emotional pains and also improves social skills. It is proven that a simple viewing of a garden benefits a person both physically and mentally.

This practice has been used as therapy for centuries. In 1798, Dr. Benjamin Rush, considered as the "Father of American Psychiatry" documented that gardening has positive effects on mentally ill patients. Following World War, II gardening was used as a form of occupational therapy for war veterans. The concept of using nature to improve human health began to gain credibility in 1970's. Rachel and Stephen Kaplan (1989) found that natural environments had relaxing and restoration benefits for people. The Journal of Therapeutic Horticulture, a peer-reviewed publication of the American Horticultural Therapy Association, has published authentic research papers since 1986. The American Horticultural Therapy Association (AHTA) serves to promote and advance horticultural therapy as a therapeutic intervention and rehabilitative approach. Degrees and certificate programs in horticultural therapy are offered at several colleges and universities and information is available on AHTA website.

Horticultural therapy is a worldwide practice and recognized as an effective treatment for clients of all ages and abilities. It is being used in several diverse settings including hospitals, rehabilitation centres, youth outreach and recovery programs, nursing homes, senior care facilities, prisons, mental health facilities, schools, botanical gardens etc. The client is engaged in horticultural activities facilitated by some trained therapists to achieve specific and documented goals. AHTA believes that horticultural therapy is an active process that occurs in the context of an established treatment plan where the process itself is considered as the therapeutic activity rather than the end product.

Horticulture therapy is a growing profession in rehabilitation and addiction centers, vocational centers, youth mental health programs, hospitals and nursing homes, and old age homes or senior citizen centers, mental health centers, schools and colleges.

Major benefits of horticulture therapy

- **Benefit from smell experience:** Studies have shown that the gaseous aromatic hydrocarbon released by plants has strong capacity of disinfection as well as infiltration against human skin, thus when it is absorbed by human skin or mucosa, it can stimulate the production of immune protein and relax the nervous system, finally enhancing the body's resistance. Recently, the report indicate that rose oil can relieve oxidative toxicity in brain caused by depression. Flowers and plants can emit different kinds of smell, which can arrive to cerebral center through 'Meatus' nerve to improve the function of the brain and stimulate the generation of pleasure.
- **Benefit from vision experience:** Visual stimulation is good for our health. The plant color, shape, size, brightness and warmness make people happy and cheerful. Red, orange, yellow, are warm colors, which are so beautiful that can make the heart beat faster. Results of laboratory research indicates that visual exposure to nature for five minutes helps to recover from stress, the blood pressure becomes normal and the tension in the muscles is released. It eases both physical and emotional pains.
- **Physical benefits**: Associating with plants, improves coordination and balance and increase muscular strength.
- **Social benefits:** Horticulture therapy in a group provides an opportunity to interact with others, instills a commitment to living things, helps learn cooperation and team working skills, teaches to deal with success and failure and provides an environment to learn and be inspired by others.
- **Emotional benefits:** Working with plants instills self-worth, confidence and self-esteem, it promotes interest and enthusiasm, satisfies the creative drive and relieves aggression and tension. Horticulture therapy also helps in cultivating positive emotions and providing a meditative mood, provides sensory simulation and mood enhancement. Mental relaxation, improvement of memory and concentration and an overall feeling of wellness. Cognitive and psychological functioning is improved. Horticulture Therapy helps to overcome stress and tension, Develop skill, inculcates self esteem and self confidence. It helps to live comfortably and independently.

- **Physical benefits:** Tending to plants provides exercise, increases ADL (Activity of Daily Living) skills, helps develop or strengthen fine and gross motor skills, increase eye-hand coordination. Working in the garden improves the muscles development, flexibility of body, coordination of mind and body, body balance and improves motor skills.
- **Physiological benefits:** Association with plants lowers blood pressure, lowers heart rate, decrease cortisol level, relieve stress.

Types of horticulture therapy

Therapeutic horticulture

The plants are considered as healers.Horticulture therapy was started with the first civilizations, when the kings used gardens to soothe and relax their minds. Monasteries and temples, where healers helped heal patients, were adorned with lovely gardens that were therapeutic in nature. Later Dr. Benjamin Rush used horticulture therapy for restoring the physical, emotional and mental health of injured and traumatized World war soldiers.

Horticulture therapy provides cognitive benefits. It enhances the attention span of the participant. Fragrance, color and feel of the plants provide peace and relaxation. It is a process through which participants strive to improve their well-being through active or passive involvement with plants and plant-related activities. In a therapeutic program, goals are not clinically defined and documented, but the leader has training in the use of horticulture as a medium for human well-being. Therapeutic benefits occur when people are exposed to fragrant flowers, green plants and participate in creative arrangement of flowers, planning, designing, sowing, planting, growing, nurturing and caring for plants for garden spaces.

Social or community horticulture

A typical community garden or garden club is a good example of a social horticulture setting. No treatment goals are defined, no therapist is present, and the focus is on social interaction and horticulture activities in a leisure or recreational atmosphere.

Vocational horticulture

It is a major component of a horticultural therapy program, focuses on providing training that enables individuals to work in horticulture industry professionally, either independently or semi-independently. These individuals may or may not have some type of disability. These programs may be found in schools, residential or rehabilitation facilities and prisons among other places.

It increases tolerance to pain and attentiveness, decreases reaction time, hospital admission time, stress, violence and health complains.

An effective horticultural therapy program should be designed:

- With good knowledge of specific populations and diagnoses.
- By devising therapeutic methods and frameworks.
- By preparing elaborate treatment plans.
- By analyzing the activities thoroughly.

25

Stress Management

Plants grow in the open in nature, hence are subjected to biotic and abiotic stress.

Biotic stress is due to living components such as:

- Insects and pests
- Nematodes, mites, snails, slugs etc.
- Vertebrates like rats, birds, jackal, bats etc.

The plant experiences biotic stress when root, stem, bark, shoots, leaves, buds, flowers and fruits are damaged by the influence and activity of biotic factors.

Biotic factors

Diseases

Like the pests, diseases also cause stress to plants. Diseases are caused by fungus, bacteria and virus. Sometimes the exposed part of insect eaten plant part becomes the entry pathway for fungus, bacteria and virus.

Fungi

Phytophthora, Pythium, Rhizoctonia sp. of fungus cause damping off, downy mildew, powdery mildew, leaf spots, fruit spots etc.

Bacteria

Bacteria cause leaf blight and root rot e.g. *Erwinia carotovora* causes root rot in banana.

Viruses

Virus causes mosaic, leaf curl and other abnormalities in plants like banana bunch top virus, cucumber mosaic virus, bhendi yellow vein mosaic virus, tomato leaf curl virus, papaya ring spot virus etc.

Other organisms that cause biotic stress to the plants are mites, mycoplasma and nematodes.

Injury caused by biotic components is by the ways, as described below:

- Injury by chewing insects- The chewing insect has teeth. The chewing insects with wings like beetles, caterpillars and grasshoppers feed on leaf tissues from leaf edge towards the center of the leaf and leaf petiole. The wingless chewing insects which crawl like the cutworms, larvae of moths, etc. are voracious feeders. They bite and chew the plant parts like leaf, root and stem, creating holes and other damage to the plant, thereby making the plant weak.
- Injury by piercing and sucking- Sucking pests like aphids and whiteflies pierce the leaves and stems to suck the sap. Large populations can cause stunting of shoots and leaf distortion like curling and yellowing. The sucking insects also produce a sticky substance called the honeydew, which becomes a substrate for sooty mould fungus.
- Damage by rasping insects- Rasping insects like mites, slugs, snails and thrips scrape off the surface of the leaves and suck the sap of the top layer of the cells and leave the leaf skeleton.
- Injury by feeding internally- Some pests remain inside the plant tissues during a part or all of their destructive stages. E.g., borers, worms, weevil, leaf miners and gall insects.
- Injury by subterranean insects- These are found in the soil and thrive by feeding on plant roots thereby causing stress to the plant. E.g., Nematodes.

Method of control

Chemical methods

- Prevention of import of plants infested with insects E.g., potato tubers with nematodes.
- Use of pesticides.
- Substance or mixture of substances used for killing, repelling or preventing insects. E.g., Monocrotophos, Profenofos (Curacron), Imidachlroprid (Confidor) etc.
- Neem oil is used to repel the pests and mostly used in "organic production" schedule.
- Use of nematicides like Carbofuran and Furadon is also common, to control soil nematodes.

Mechanical methods

- Insect population is directly killed by mechanical devices.
- Insect traps are used to trap fruit flies.

- Light traps are used to trap moths and beetles.
- Hand picking of egg masses and larvae.
- Wrapping the fruits with butter cover or muslin cloth while the fruit is still on the plant e.g., Pomegranate fruit borer.

Cultural methods

- Adoptation of cultural practices like deep ploughing which eliminate or reduce the insect population.
- Planting trap crops to trap the insects so that they do not harm the main commercial crop e.g., planting mustard in every 25th row of cabbage to trap diamond back moth.
- Following crop rotation to reduce the insect population e.g., planting banana after papaya.

Biological control

This refers to controlling the pests with the help of another living organisms. Fungus, parasitoids and predators are used.

- It is less expensive and offers a long-term control.
- No pollution and no risk to human health.
- Bio control agents are abundantly available in nature.
- Some fungus attacks the pest by making colonies on the pest e.g. *Bacillus thuringiensis and Metarhizium anisopliae*
- Parasitoids are the organism which complete their life cycle in the body of the host and ultimately kills ite.g., *Acerophagus papayae* (Papaya mealy bug).
- Predators are the insects which feed on other harmful insects. Predators are harmless to the crop. e.g. Lady bird beetle (*Coccinella* sp.) predates on several sucking pests.

Abiotic factors

Temperature

Optimum temperature is important for the best performance of the plant at all its developmental stages. Important functions like photosynthesis, respiration, water absorption and transpiration require a particular temperature. Response of plant to extremes of temperature varies. When the temperature goes below 10°C 'choke throat' in banana is seen. Other impact on plants at a very low temperatures is chilling injury where ice crystals are formed and mechanical injury is caused to the plants. On the other hand, temperature higher than optimum causes wilting and desiccation due to excessive water loss.

Light

Intensity of light, quality of light and light duration affects germination, tuber, bulb formation and sex expression. In cocurbits, long day condition promotes the development of male flowers.

Atmospheric humidity

High atmospheric humidity increases the incidence of pests and diseases.

Rainfall

Total rainfall and distribution affect the growth of crops. Continuous heavy rainfall for a period of 2-3 days during peak flowering period washes off the pollen, hinders insect pollination and dilutes the stigmatic fluid. Timely receipt of summer showers in February-March decide flowering and fruiting in coffee. Rains during harvest causes rotting in grapes. Optimum rainfall for some crops is given below.

	Crop	Rainfall (cm) / year
1.	Mango	25 – 250
2.	Pepper	125 – 200
3.	Cardamom	200 – 250
4.	Rubber	200 – 250
5.	Dates	15 – 25

Wind and hailstorms

Heavy wind causes physical damages by shedding of flowers, fruits, breaking branches and uprooting trees. Frequent dry winds cause desiccation and in such a case, irrigation is required at frequent intervals. Hail storms during pre-blooming or blooming period of apple, plum, peach affect fruit set.

Air pollutants

Air pollutants like ozone, sulphur-di-oxide, oxides of nitrogen, reduce the assimilation rate of plant and reduce the yield. In Punjab, UP, Bihar and West Bengal, "Black tip" a disorder in mango occurs when the brick kilns are within the radius of 1.5 km. Brick kilns emit carbon di-oxide, sulphur dioxide and acetylene gases, which leads to black tip.

Frost

It is a thin layer of ice crystals deposited on soil and plant surfaces as a result of freezing temperature. Frost occurs in hill stations, during the winter months.

Orchards at the bottom of the slope or valley in Nilgiris, Valparai, Yercaud and Kodaikanal experience frost. Frost damages the potato, pea and cole crops in South Indian hills.

Altitude

Altitude affects temperature. For every 100 m increase in elevation above MSL, a decrease in temperature by 1-2°C is noticed. Crops which can adjust to high altitude conditions should be grown, wrong choice would cause stress to the plants and the plant performance would be affected. For example, tea and coffee perform without stress from 1000 to 1800m above MSL. Similarly hill banana has a poor taste when grown in plains. Tropical fruits do well at 1000 m below MSL. Coconut, if planted at 1000 to 1200 m MSL, takes 10-12 years for flowering.

26

Post Harvest Technology

Horticultural crops are highly perishable and the post-harvest loss is estimated to be 20-30% every year. There are various factors associated with post-harvest losses. It includes pre-harvest cultural operations, maturity index at which the produce is harvested, methods of harvest, pre and post-harvest treatments, packing materials and storage methods. Postharvest technology can be defined as the techniques which help to reduce the post-harvest losses.

Assessment of harvesting stage

Climate may be related to periodical events in plants life. The science dealing with this relationship is termed as phenology. The phenological data for horticultural crops include dates of planting, germination, flowering, fruit set, ripening and harvest. These dates are determined by the climatic factors prior to as well at the time of harvest. A useful phonological concept for the prediction of the harvest time is based on degree days. A " Degree day" can be defined as a day when mean daily temperature is one degree above the minimum temperature for the growth of the plant. The minimum temperature for the growth of a particular plant is termed as "zero temperature". Zero temperature vary according to crop. E.g. if zero temperature of a particular crop is 50°F, then a day with mean temperature of 68°F would provide 18° days. The degree days are also called as "heat units". It can be determined by how many degree days are needed for the horticultural crop to reach harvesting stage.

Harvesting

It is one of the crucial features in horticultural operation. It is often the costliest operation. The timing of harvest has a direct bearing on the final quality of the product. Often, there are a number of conflicting factors to be considered with regard to time of harvest. For examples, in apple, the storage quality is affected by delayed harvest. But, red colour development tends to increase with time. Determining the optimum time of harvest is not a single and straight forward decision. Factors affecting the date of harvest are:

- The genetic nature of cultivar or variety
- Planting date or blooming date
- Environmental factors

There are different cultivars mature at different dates. Therefore, the planting of assortment of such cultivars can make for an orderly succession of crops throughout season ensuring regular supply to the market. In some crops, the time required to reach the harvesting stage may be expressed with temperature-time value called "heat units". But following only the heat units as a measure to judge the time of harvest has some limitations.

- The soil temperature more accurate than air temperature
- Difference in day, night temperature shifts
- Differential effect of temperature above a minimum may not have a similar effect on growth

Hence, there are various physical, chemical and biochemical methods are developed to ensure the optimum stage and time of harvest.

Maturity

Maturity can be defined as state of ripeness. It is defined as the stage at which a commodity has reached a sufficient stage of development and ready for harvesting or Maturation is the stage of development leading to the attainment of physiological maturity (when a plant or plant parts will continue ontogeny even if detached) or horticultural maturity (when a plant or plant parts possesses the pre-requisites for utilization by the consumer for a particular purpose. The stage of development results in the maximum quality of the product. Maturation is indicative of the fruit being ready for harvest. It is determined by various factors. Different standard of maturity is required for different uses. It depends on the product which is picked for local market, distant market, shipment and air freight. The edible part of the fruit or vegetable may be fully developed in size, although it may not be ready for immediate consumption. Ripening follows or overlaps maturation, rendering the produce edible, as indicated by taste. Senescence is the last stage, characterized by natural degradation of the fruit or vegetable, as in loss of texture, flavour, etc.

Maturity indices

Maturity indices are important for deciding when a given commodity should be harvested to provide some marketing flexibility and to ensure the attainment of acceptable eating quality to the consumer. Maturity index for a commodity is a measurement or measurements that can be used to determine whether a particular commodity is attained maturity.

The indices are important to trade regulation, marketing strategy and for the efficient use of labour and resources. For maturity, measurements to be carried out by producers, handlers and quality control personnel, must be simple, readily performed in field or inspection point, and should require relatively inexpensive equipment. Index should preferably be objective (a measurement) rather than subjective (an evaluation) and ideally index should be non-destructive.

Determination of harvest maturity indices

There are many methods to determine the harvest maturity. But some methods are practiced frequently. They are as follows,

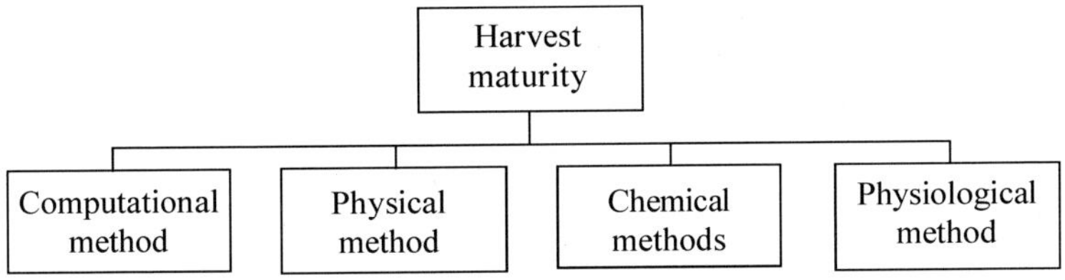

I. Computational methods

1. Calendar date
2. Days from full bloom to harvest
3. Mean heat units
4. T-stage

II. Physical methods

1. Fruit retention strength
2. Fruit size and surface morphology
3. Weight
4. Specific gravity
5. Colour (skin, flesh and seed)
6. Flesh firmness

III. Chemical methods

1. Total soluble solids (TSS)
2. Titrable acidity
3. TSS/acid ratio
4. Sugars (total and reducing)
5. Sugar/acid ratio

6. Starch content
7. Starch-Iodine test
8. Tannin content
9. Oil content
10. Juice content

IV. Physiological methods

1. Respiration rate
2. Ethylene evolution rate

I. Computational method

1. Calendar date

Calendar date is one of the most commonly used indices of maturity and reasonably accurate if factors of time of bloom and weather during growing and harvesting season is considered.

2. Days from full bloom to harvest (DFFB)

DFFB may provide a better index of maturity than calendar date provided number of days used has been derived in the district where it is being used as an index. DFFB is most reliable index for fruit crops. However, it varies greatly from year to year and from location to location due to environmental fluctuations.

3. Mean heat units

The apparent effect of temperature on flowering and fruit development suggest use of degree days or heat summation for more accurate prediction of harvest date. Accumulated heat units in apple for optimum maturity are computed by calculating time in relation to temperature above a certain minimum.

4. T- stage

Another method related to maturity is the date when fruit reaches T-stage. It is a point at which the angle formed by fruit receptacle and pedicle reaches 90° and its timing is obtained by interpolation. Angle between pedicle and receptacle is greater than 90° early in growing season and less than 90° in late season. Number of days between T-stage and date of optimal harvesting time of apple showed a remarkable consistency for a given location.

II. Physical methods

1. Fruit retention strength

The retention strength of the fruit is higher when the fruit is in immature stage. When a fruit is in mature harvestable stage, the retention strength is weaker than in immature stage.

2. Fruit size and surface

Fruit size is used as a maturity index in many crops like capsicum, okra etc. Changes in the size of a crop while growing are frequently used to determine the time of harvest. Fruit size depends upon a number of variables such as soil, cultivar, fertilizer application and climatic conditions which may vary from place to place and season to season.

3. Weight

Weight is poor measure of maturity indices. It mainly depends on number of variables such as soil, cultivar, fertilizer application and climatic conditions which vary from place to place and season to season.

4. Specific gravity

Specific gravity is another measure of maturity, as vegetables or fruit matures specific gravity increases. Specific gravity is obtained by comparing the weights of equal bulks of other bodies with the weight of water. In practice, the vegetable is weighed in air, then in pure water. The weight in air divided by the weight in water gives the specific gravity. In this method, fruit or vegetable is placed in a tank of water, where immature fruits float and mature ones sink.

5. Colour (skin, flesh and seed)

This factor is commonly applied to fruits, since skin colour changes as fruit ripens or matures. Some fruits exhibit no perceptible colour change during maturation, depending on the type of vegetable. Assessment of harvest maturity by skin colour depends on the judgment of the grower, but colour charts are available for cultivars, such as apples, tomatoes, chilli, etc. For scientific observation, digitl chromometers are available to measure the intensity of colour in horticultural crops.

6. Flesh firmness

A fruit may change in texture during maturation, especially during ripening when it may become rapidly softer. Excessive loss of moisture and ripening may also affect the texture of crops. These textural changes are detected by touching the fruit. The grower may simply be able to gently press and squeeze the fruit and judge whether the crop can be harvested or else representative samples may taken from field and tested in texture analyzer or tenderometer.

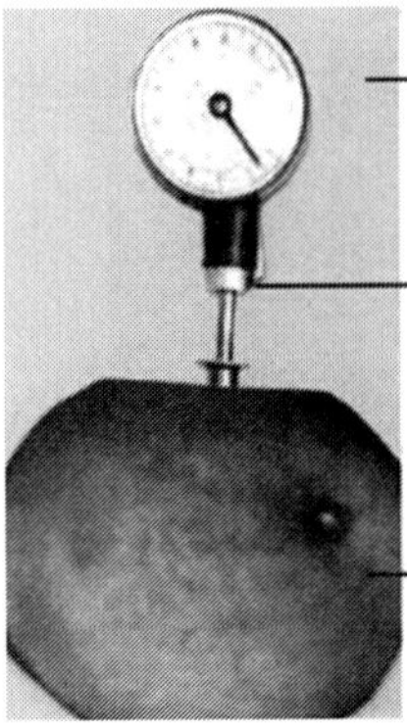

Pressure tester to measure firmness

III. Chemical methods

Common chemical methods are used to determine the harvest maturity of certain fruits and vegetables. They are as follows,

1. Titrable acidity

In many fruits and vegetables the acidity changes during maturation and ripening. Take samples of fruits and vegetables and extract the juice. Then it is titrated against a standard alkaline solution, gives a measure that can be related to optimum time of harvest. Normally, acidity is not taken as a measurement of fruit maturity by itself, but in relation to soluble solids, which is termed the **brix: acid ratio.**

2. TSS/ acid ratio

Soluble Solids content of fruit is also determined by passing light on the fruit or vegetable and measuring the amount transmitted. This is a laboratory technique, done by refractometers and might not be suitable for village level production. A drop of fruit juice is placed in the sample holder of the refractometer and a reading taken; this is equivalent to the total amount of soluble solids or sugar content.

3. Sugars (Total and reducing)

Majority of soluble solids present in juice is contributed by sugars, which give a reliable measure of sugar content. A quick method to measure the amount of sugar present in fruits is with a ***brix hydrometer or a refractometer.***

Otherwise, total and reducing sugars can be tested in laboratory analysis.

4. Sugar /acid ratio

Sugar acid ratio decides the sugar acid blend in fruits, which is the quality decides the taste.

5. Starch content

Measurement of starch content by calorimetric method may help to determine the optimum maturity for harvest as a text may help.

6. Starch- Iodine test

Cut the fruit in two pieces and dip the cut pieces into a solution containing 4% potassium iodide and 1% iodine. After one minute, fruits are removed and treated surface is rinsed with running tap water. The cut surfaces stained with blue-black colour indicates the presence of starch which can be read and photographed immediately. Starch converts into sugar towards maturity. Harvest begins when the samples show 65-70% of the cut surfaces have turned blue-black.

7. Tannin content

8. Oil content

Oil content which increased with the onset of maturity can be used as a reliable index for maturity. This method is more successful for cultivars naturally high in oil content. For example in Avocado, fruit oil content should not be less than 8 % by weight excluding skin and seed.

9. Juice content

The juice content of many fruits and vegetables increases as it matures on the tree. To measure the juice content of a fruit, a representative sample of fruit is taken and then the juice extracted in a standard and specified manner. The juice volume is related to the original mass of juice, which is proportional to its maturity.

IV. Physiological methods

Various physiological methods used for determining horticultural maturity are as follows.

1. Respiration rate

Respiration rate is calculated by measuring the uptake of oxygen or carbon-di-oxide evolution from fruit. When climacteric fruits or vegetables reach physiological maturity and ripening processes are initiated, there is a marked increase in respiratory activity.

2. Ethylene evolution rate

Ethylene evolution rate is measured using ***gas chromatography*** mostly in climacteric fruits or vegetables. The fruits must be harvested at full maturity with slight colour development on the beak. However, low ethylene production does not necessarily mean that fruits possess good eating quality after storage.

V. Other methods

1. Optical methods

Light transmission properties can be used to measure the degree of maturity of vegetables. These methods are based on the chlorophyll content of the vegetables and fruit, which is reduced during maturation. The vegetable or fruit is exposed to a bright light, then switched off, so that the vegetable or fruit is in total darkness. Next, a sensor measures the amount of light emitted from the vegetable or fruit, which is proportional to its chlorophyll content and thus its maturity is assessed put this technology may be applicable for coloured fruits and vegetables.

2. Leaf changes

Leaf quality and colour change often determines when fruits and vegetables should be harvested. For example, in potato, drying of leaves and stems is the indication for attainment of maturity. If harvested earlier, the skins will be less resistant to harvesting and handling damage and more prone to storage diseases.

3. Abscission

As part of the natural development of a fruit an abscission layer is formed in the pedicel. For example, in cantaloupe melons, harvesting before the abscission layer is fully developed results in inferior flavored fruit, compared to those left on the vine till the abscission layer is furmed.

Maturity indices for horticultural crops

- Colour change is one of the major criteria in many fruit and vegetable crops. In mango, change of peel colour from green to yellow and in papaya change of colour at apical end of fruit. In pumpkin, change of colour from green to brown colour and in ash gourd disappearance of ashy bloom on the surface.
- Increase in size, attaining optimum size which can be predicted based on the experience of a farmer or technical person working in the field. Size vary with different varieties and cultivars.
- Softening of the tissue of the fruit e.g., figs and sapota
- Ease of detachment from the stalk e.g., sapota, annona
- Reduction in latex exudation e.g., sapota and papaya
- Shriveling of fruit stalk e.g., water melon0

- Time elapsed from the date of flowering to picking maturity
- Sound by tapping- jack and water melon when ripe produce hollow and dull sound on tapping but produce metallic sound if unripe
- Many of the root vegetables like tapioca, turmeric, carrot, beet root drying of lower leaves and cracking in the soil around the base of the stem is noticed.

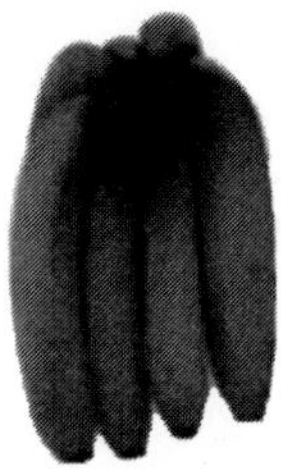

1. All green

2. Green with a trace of yellow

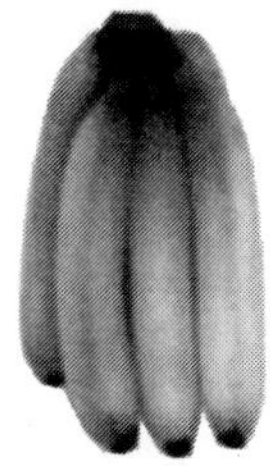

3. More green than yellow

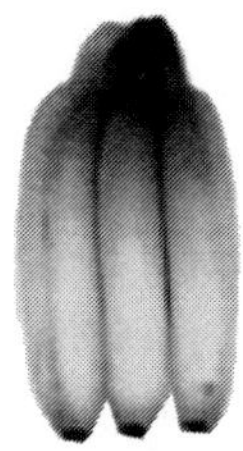

4. More yellow than green

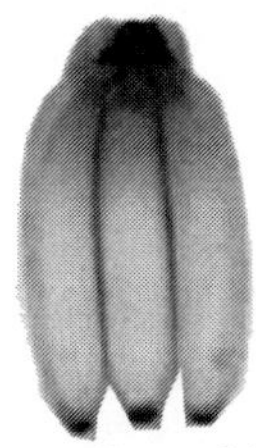

5. Yellow with a Trace or Green

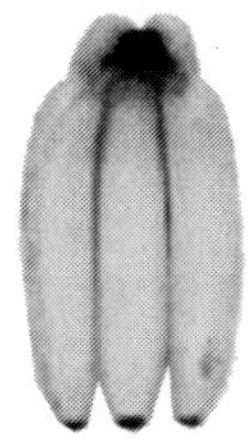

6. All yellow

7. All yellow with brown speckles

Harvesting methods

Specific harvest procedures is advocated for each crop but there are some general handling and harvest practices common to all produce. Produce should be handled with care to minimize mechanical damage. Mechanical damage renders the produce unmarketable and allows for disease organisms to enter. Some vegetables are more durable than others, but all should be handled gently. Handling and harvest containers and utensils should be cleaned and disinfected daily. This will reduce postharvest losses due to disease. Cutting utensils should be sharp to prevent damage to the plant. Workers should be trained on how to use sharp knives and shears safely to prevent injuries. Transport vehicles and field handling areas should be cleaned after harvest daily.

Harvesting methods

Hand harvesting: It is still now used practical method of harvest for many high value crops that are either sensitive to bruising or selectively picked, crops.

Completely mechanized: All operations are performed by a single machine in a single step. E.g., Potato. One single machine that digs and removes the vine and loads the tuber

Semi-mechanized: Fruits are picked in hands and put on the conveyors immediately when they are picked in the field and moved to factories. Cranberries were the first fruit crop harvested by machines. The trees are essentially combed to remove the berries. Equipments have been developed for shaking plums, peaches for harvest. Tractors fitted with small equipments can move around orchards to harvest the fruits very safely.

Pre marketing operations

- **Washing**: It is done for enhancing appearance and removal of any dirts on the surface of the produce.
- **Trimming**: It is done to cut back discoloured leaves and green tops.
- **Waxing:** It is done to improve the appearance and slow down the moisture loss which causes shriveling.
- **Curing**: In the case of root and tuber vegetables, curing helps in reducing post-harvest decay and water loss at farm level at little or no additional cost. It is done immediately after harvest. It strengthens the skin of tubers. Curing of onion includes drying of superficial leaves of onion bulbs which protect them from microbial infection in storage and also gives attractive skin colour to onions. Maximum safe temperature for onion curing at field level is found to be 37.8°C for 6 days. Artificial curing of onion in crates at 40°C for 16 hours is found to reduce losses due to rotting in storage. In other crops like potato, sweet potato, garlic, cassava, yam, curing effectively reduces post-harvest decay and water losses. Potato curing is most effective at about 20°C with atleast 85% RH, whereas sweet potatoes are effectively cured at 29°C.
- **Pre cooling**: This is done for rapid removal of heat from the recently harvested fruits. When the vegetables are harvested during hot weather, to remove the field heat, to retard the ripening and senescence, pre cooling is necessary. When harvested during hot period, pre-cooling of melons is of vital importance. It slows down the rate of respiration. It reduces wilting, shriveling by decreasing the water loss. It inhibits the growth of decay causing micro organisms. Precooling is an energy requiring process. The cost can be minimized if the harvest is done during night hours or early morning. Different methods of precooling are hydro cooling, contact icing, vacuum cooling and air cooling etc.

- Placing the produce in refrigerated trucks with forced humidified air circulation.
- Placing ice in package
- Placing ice in water and passing the produce through a spray of cold water

Hydro cooling and vacuum cooling are the two methods of storage of vegetables. Hydro cooling is accomplished by flooding; spraying of water or immersing in water is a rapid and effective method because water is an excellent material to transfer heat from the produce to the cooling medium. A minute quantity of fungicides can also be added in hydro cooling water to control fungal growth. In case of large sized vegetables like brinjal slow cooling occurs because of high surface: volume ratio.

Highly perishable leafy vegetables and salads are often cooled by vacuum cooling method by reducing the atmospheric pressure in hermetically sealed chamber until the reduced vaporizing point of water produced at low pressure in the cooling chamber cools the produce. This is called vacuum cooling. The advantage of this method is speedy and uniformity in cooling of the commodity. Leafy vegetables are difficult to cool rapidly and uniformly by vacuum cooling.

- **Grading**: It is done for uniformity in quality, shape, size, colour and ripeness. The high graded material only fetch premium price in the market.
- **Prepacking**: The produces are placed in bags of transparent plastic films or in trays covered with transparent films. The films may be of polyethylene materials with permeability to oxygen and carbon-di-oxide

PACKAGING: The use of proper package material for transport and marketing of vegetables can reduce their losses and maintain their freshness, succulence, and quality for longer periods. Improper packing material and inconvenient size of the material deteriorate the quality and appearance. In India, vegetable marketing is not developed fully as in other countries. Many vegetables are transported in gunny bags or bamboo baskets. Cheap packaging materials like polyethylene film, paperboard boxes lined with polyethylene and other materials can effectively prolong the shelf life. Vegetables can be protected from air by packing in various types of plastic films. Polyethylene provides modified atmosphere and consequently reduces decay, softening and loss of total soluble solids as in tomato, bhendi, brinjal and cauliflower. The thickness and permeability to carbon dioxide and oxygen and water vapour of the film needs to be standardized for vegetables. For e.g., brinjal and cauliflower needs thin films otherwise it develops off-flavour. Thin films maintain flavour of cauliflower for longer period. Bhendi can be stored for 9 days without reduction in quality. Bagging of carrots in perforated films significantly reduces weight and water loss in transportation.

Crop	Type of package
Lettuce	Paperboard lined with polyethylene
Cabbage (individual), cauliflower	Polyethylene
Spinach	Transparent films
Tomato	Upright cone baskets with grass in between two layers

Package material may be rigid or flexible. Flexible types are used in low temperature storage.

Storage

Low Temperature Storage: Refrigeration or low temperature storage can most effectively extend shelf-life of vegetables and reduce post-harvest losses by arresting metabolic breakdown and fungal deterioration of the commodity. Many tropical and subtropical vegetables are susceptible to low temperature injury. Lack of capital to establish and run the cold storages individually limits the use of cold storage in developing tropical countries like India. The rate of respiration has direct correlation with temperature. Higher the temperature, more is the rate of respiration and multiplication of decay organisms. Low temperature, Relative Humidity (RH) of the cold storage is another important factor. Water loss from vegetable is rapid at low RH because air in the room contains less water vapour than it can hold at the temperature of the room. The vegetables are saturated with water and water vapour is readily transferred from the humid interior of the vegetable to the relatively dry air. At higher RH, microbial growth takes place and therefore, appropriate percentage of RH is important. For most of the vegetables, storage temperature just above freezing point is desirable.

Environmental control: Fresh vegetables consume oxygen for respiration and give out carbon dioxide and emit ethylene. The emitted ethylene further speeds up the ripening of vegetables or yellowing of leafy vegetables. Changing the micro climate around stored vegetables, reducing O_2 and increasing CO_2 and removing ethylene can prolong their shelf life. These conditions can be full filled by "controlled environmental storage". It includes

- Controlled atmosphere storage (CAS)
- Modified atmosphere storage (MAS)
- Hypobaric storage

Both, CAS and MAS involve manipulation of CO_2, O_2 and N_2 levels of storage rooms. Other gases like CO_2, C_2H_4, acetylene, propylene are also considered. MAS differ from CAS only in how precisely gas pressures are modified. CAS is more precise than MAS. Hybobaric storage is a type of CA storage with emphasis on reducing the atmospheric gaseous pressure exerted on the stored material.

Controlled Atmospheric Storage (CAS): In this method of storage, higher CO_2 and lower O_2 are maintained. If it is combined with refrigeration, it retards respiratory activity and may delay softening, yellowing, quality changes and other deteriorative process. The important factor in CAS is tolerance or susceptibility of the vegetable to the injury caused by decreased oxygen and increased carbon dioxide concentrations. The relative tolerance of the vegetable varies considerably and it depends on the temperature, age, morphology and anatomy of the vegetables. The effect of CAS on respiration depends on the type of vegetable and concentration gradient that develops between the centers of metabolic activities and outer integument of the material. The combined effects of reduced O_2 (2%) and increased CO_2 (1%) result in reduction of respiration rate by 50%. The advantages of CAS are:

1. Prolonged shelf life of perishables by arresting respiration and senescence
2. Reduction in produce sensitivity to ethylene
3. Decrease in incidence and severity of decay and control of fungai, bacteria, insects and pests in selected vegetables.
4. The CA storage is beneficial for vegetables that deteriorate rapidly or those that complete ripening after harvest. For example 1% O_2 was very effective in retaining ascorbic acid content of chinese cabbage in delaying losses in sugars and chlorophyll contents and in extending the storage life at 0 °C.
5. Elevated CO_2 (5-20%) in CA storage reduces severity of chilling injury symptom in several vegetables including okra, chilli pepper.

The CAS has its demerits

1. Aggravation of physiological disorders
2. Irregular ripening
3. Development of off flavour
4. Accumulation of organic acids
5. Unfavourable changes may occur due to induced high CO_2 and low O_2 concentrations.
6. A fixed levl of gases may not be suitable for all circumstances because the temperature and duration of storage and the commodity to be stored are inter-dependent factors, each influencing the use of gas in CA storage.

Post Harvest Treatments

1. **Fungicides:** The control of micro-organisms by pre-harvest spray of fungicides result in reduction in the load of pathogens on the harvested produce. The best method of application of fungicides and antibiotics is to mix them in wax emulsions. Fumigation of stored material with sulphur dioxide prevents spread of fungi like *Botrytis*. Biphenyl impregnated tissue wrappers control *Penicillium*. Benomyl prevents growth of most damaging fungi of tomato. Dipping in 1000 ppm of streptomycin delayed the bacterial soft rot of packed lettuce and spinach.

2. **Anti- transpirants:** Prevention of water loss in harvested vegetables in general and leafy vegetables in particular is one of the important aspect. For this purpose, wax emulsions and film packaging are used. The use of anti-transpirants that particularly close the stomata would give additional benefit in moisture conservation, especially in leafy greens.

3. **Growth substances:** Increased membrane permeability, respiration, phosporylation and shift in protein synthesis lead to deteriorative processes and eventually senescence. As long as the produce is attached to the metabolic supply, it continues to receive the plant growth hormones, especially auxins, gibberellins and cytokinins. The effect of cytokinin is more predominant in leafy vegetables. Eg., N 6 - benzyl adenine (BA) is one of such type used at the concentration of 5-15 ppm. Post-harvest application of gibberellins retard ripening in tomato. Pods of auxin treated plants remained more green even after those from non treated plants. Application of 2,4 D along with BA retard yellowing of cauliflower. Pre- harvest spray of Naleic hydrazide (MH) inhibit sprouting in onion and potato. Application of Alar inhibit the synthesis of solanine in potato. BA at 50mg/lit was found to be most effective in reducing decay loss and maintaining quality parameters of ridge gourd at 0.5°C.

4. **Ethylene absorbants:** Ethylene (C_2H_2) is usually emitted by senescing tissue and it acts as an autocatalystic of senescence. Limiting the production of C_2H_2 by tissue, making the environment of storage unfavorable for C_2H_2 action and by fast removal of emitted C_2H_2 around the produce, senescence process can be delayed and prolong shelf life of the commodity. Several chemical formulations such as brominated activated carbon, celite with KMnO4, KMnO4 on vermiculite, Purafil R- alkaline KMnO4 on a silicate carrier in packages are used successfully to absorb ethylene evolved by the produce. Application of KMnO4 1000 ppm on paper lining which is a used as cushioning material for packing tomato fruits, enhances shelf life of fruits and decreases weight loss.

5. **Calcium:** It has desirable effects in delaying senescence and controlling physiological disorders in vegetables. The rate of senescence often depends on the calcium status of the tissue and by increasing calcium levels, various parameters of senescence such as respiration, protein content, chlorophyll content and membrane fluidity are altered. Various calcium treatment methods are pre harvest spray and post harvest dipping.
6. **Ionizing radiation**: A useful method. It destroys the pathogenic organisms without much change in temperature of the produce. The initial investment is high but it is easy for handling and it leaves no residues and need minimal energy. Ionizing radiations extend the shelf life of vegetables by surface pasteurization, sprout inhibition, and retardation of senescence process. Ionizing radiation inhibits sprouting of onion and potato effectively at 10-15 krads. Irradiation of potatoes may damage the tissues which can be eliminated during the storage by some treatments. The irradiation dose above 0.15 kGy may cause undesirable effects like decreased wound healing ability, tissue darkening, increased sugar content in potatoes, decreased vitamin content and increased susceptibility to post harvest pathogens.

   Anthocyanins and lycopene are more susceptible to degradation by radiation at doses higher than 1.0 mega rad than carotenoids and chlorophyll pigments. Vitamin C is found to be sensitive than thiamine, riboflavin and pyrodoxin by sterilizing dose of X rays and cathode rays. The fruit and vegetable softening is associated with degradation of proteins, fat and carbohydrate into simple compounds. These changes could be arrested by irradiation at 2-3 x 10^5 rads. Taste preference of irradiated asparagus, string bean and cabbage is good after 10 days of storage. The possible use of ionizing radiation for insect disinfestations is one of the most promising applications.
7. **Inpackage desicants:** To maintain the succulence and tenderness of the vegetables, partial dehydration is done and then in-package desiccants are used to limit the moisture availability to the microorganisms. The use of in package desiccants such as calcium oxide or fumed silica's to bring the moisture content of the storage environment to 1% or lower permits the storage of dehydrated vegetables for longer period.

Transportation: In India, the transport of the vegetables is by bullock cart or tractor trolleys. The long distance transportation is mainly by road and it is very costly. Only 20% of the total produce is transported by mechanical means. The contribution of rail transport is relatively low in India. The losses in transportation occur due to physical and mechanical injuries. Under uncontrolled transportation conditions, mainly temperature and relative humidity, losses of huge magnitude may occur. For long distance transport, especially by rail the produce should be properly packed and staked in the carriages. The carriages should be ventilated.

27

Processing, Preservation and Value Addition

India is bestowed with varied climatic conditions and a variety of horticultural crops are produced. It is estimated that 20-30% of the produce is lost before consumption because of various reasons including improper storage and this is where the need of processing for preservation arises. There are various ways of preserving the produce and various ways to process it.

Preservation is defined as methods of extending the shelf life of the produce in its original form or in a different usable form. The basic principle of preservation involves strategies which aim at reducing the spoilage. The horticulture produce is spoiled by microbial infection or infestation. Prevention of microbial damage can be achieved by keeping the produce away from microbes, by removing the microbes, obstructing the proliferation and by killing the microbes. Infestation in the harvested fruits may lead to other microbial infection and spoilage due to rotting.

1. Preservation by dehydration

- The practice of drying fruits, vegetables, herbs and flowers is an age old practice. Moisture is removed from fresh produce by exposing the produce to intense sun light. The dehydrated products shrink to about one tenth of the fresh produce, so can be stored in less space and also are easy to transport. Fruits like mango, banana, grape, fig, jack fruit, sapota, pineapple, amla, peaches, plums, etc. are sun dried and used.
- Vegetables like moringa, slices of bittergourd, potato, carrots, radish etc. are sundried and used in off season. Moringa leaves are dried and powdered consumed for health benefits. Capsiums, carrots and tomaotoes are diced and dried and used in soup. Tomato pulp is dried and powdered and used for various culinary purposes. Spices like chilli, coriander, turmeric are sun dried and used. Flowers like rose, jasmine are dried and used in potpourris. Basil, rosemary, thyme, oregano, mint leaves are dried and stored to be used when required. Now, due to advancement in technology, dehydration is carried out by machines in absolute hygienic conditions.

2. Preservation by heat

- **Pasteurization**- The heating temperature is maintained below 100°C. It is done by steam, hot water or dry heat. The products when packed in containers, pouches, hot packs or hot fills need to be pasteurized.
- **Blanching**- Temperature is maintained at 100°C. The produce is dipped in hot water for a short span of time and dried or frozen for example, frozen peas, tomatoes, sauerkraut.
- **Sterilization**- The temperature is maintained above 100°C. Vegetables and fruits are heated continuously for 30-90 minutes above 100 degree centigrade. The cans and bottles used for preservation should be sterilized before filling.

3. Preservation in low temperature

- **Underground cellar storage** – Root and tuber crops are stored in these structures during colder months of the year. This is a method to store for a short duration.
- **Refrigerated storage** - The storage of the produce is done for a short duration in zero degree centigrade.
- **Slow Freezing** - Freezing to – 15-29°C slowly in about 72 hours.
- **Quick Freezing** - Freezing to 0 to 4°C in less than thirty minutes. This is done by direct immersion in brine or sugar or indirect contact with refrigerant or by air blast method.
- **Cryofreezing**- It is the method where the produce is stored below -60°C with the help of liquid nitrogen and liquid carbon dioxide.
- **Dehydrofreezing** – Freezing is done after partial dehydration, i.e., about 50 per cent of natural moisture is removed.

4. Freeze drying

The product is frozen at -18°C.

5. Preservation by chemicals

Preservatives retards or arrests the growth of microbes. In pasteurized products like squash crush, cordial, RTS, nectar, jam jelly, marmalade, sauces, ketchup etc. potassium meta-bi sulphate, citric acid or benzoic acid are used.

6. Preservation by fermentation

Natural fermentation by microbes to produce alcohol and vinegars is important natural way of preservation of fruits and vegetables. Examples are beer, wine, fermented beverages and pickles.

Minimal processing

Unlike olden days, the urban population does not have time to do all the chores in the kitchen and vegetable washing and cutting being one of them. So, in present days, minimal processing is a technique where the fruits and vegetables are washed, trimmed, cut and packed in for ready use. The minimal processed fruits and vegetables retain the fresh quality of the fresh fruit, and more shelf life than the fresh fruit owing to hygienic packaging and storage. Since minimal processing does not use any dehydration or heating process or any chemicals the nutrient quality, flavour and aroma are retained. Minimally processed fruits and vegetables are prepared by washing, cutting, grating, shredding, removing the leaves etc. and packing at chilling temperatures under polymeric films which provide optimum modified atmosphere packing (MAP) conditions. This provides good choice to consumers, who perceive good fruits and vegetables as a healthier choice over the processed alternatives.

The following choices are available as minimally packaged foods

1. Cut mangoes, pine apple, papaya, kiwi, dragon fruit, pomegranate either individual packs or as mixture of fruits.
2. Vegetables like banana stem, banana flower, finely cut beans, grated carrots and diced capsicums.
3. Moringa leaves and flowers detached from petioles and packed in polythene bags with perforations are available.

Value addition

1. **Jams:** Jam consists of 45% fruit pulp and 55% sugar. The preservative is 0.5-0.6%. Jam is made up of fruit pulp of sapota, papaya, mango, grape, guava, apple, peach, pear, plum etc. Jams are packed in bottles with adequate sealing technology in order to increase its shelf- life.
2. **Jelly:** It is a semi-solid product prepared by boiling the fruit to remove the pectin. It is free of pulp, having 55% solids, 45% fruit extract and 0.5- 0.7% acid. Guava, papaya, apple, karonda are fruits rich in pectin, thus rendering them perfect for jelly-making. Jellies are also packed aseptically by vacuum by glass bottles.
3. **Marmalade:** It is a fruit jelly with suspended fruit pieces and peel, usually that of citrus fruits.
4. **Candy:** Fruits are immersed in cane-sugar, drained and dried to attain this preserve, fruits that are often candied are amla, pineapple, karonda, citrus peels, cherry etc. candy has 70% sugar.

5. **Crystallised fruit:** The fruits, once candied, are coated with sugar crystal, either by sprinkling the sugar, or allowing the dense sugar syrup to form crystals on them. The candied fruits are drenched in 70% sugar solution and left undisturbed for about 18 hours, then are dried in driers at 50°C.

6. **Glazed fruit:** The candied fruits are coated with a thin layer of sugar syrup, giving it a glossy appearance.

7. **Chutneys:** Chutneys are made up of ripe or raw fruit or vegetables. It is either grated or cut. It is cooked for a long time below the boiling point, with spices and vinegar. Chutneys are usually made of mango, tomato, tamarind etc.

8. **Preservation in brine:** Preserving vegetables in 2% salt solution and addition of vinegar and sugar is also done in some cases. Gherkins, tender cucumbers are preserved in brine solution.

9. **Sauerkraut:** Finely cut raw cabbage preserved in brine. It has been fermented by various lactic acid bacteria. It has a long shelf life and a distinctive sour flavor, both of which result from the lactic acid formed, when the bacteria ferment the sugars in the cabbage leaves.

10. **Pickles:** Fruits or vegetables are preserved in common salt or vinegar. This process is called pickling. They are made of mangos, tomatoes, chilli, garlic. Oil or vinegar is added to preserve pickles.

11. **Sauces and ketchups:** Sauces are mixed with fruit pulp and spices, and are thinner than ketchup. The former is made up of papaya, tomato, apple etc. It has 25% TSS and 1% acidity. While sauces have filtered ingredients, the latter has fruit pulp in it, thus making it thicker than sauces.

12. **Puree and paste:** Tomato puree is most common and has 9% salt with 12% tomato solids. If the solids reach upto 25%, it becomes tomato paste.

13. **Syrup**: Syrup is a fruit beverage containing 65% TSS, 25% juice, 1.5% acidity and 600 ppm Potassium Metabisulphate (KMS). Fruits such as amla, jamun, grape and citrus fruit are used to make syrup.

14. **Fruit juices:** Fully ripe fruits are selected to extract juice. It is extracted, filtered, clarified and mixed with sugar. Sometimes fortification of juice is done with vitamins. Juices are sometimes bottled in their natural form as well, without addition of sugar.

15. **Ready To Serve (RTS):** This type of fruit beverage has got 10% juice, 10% solid and 0.3% acid. It is not diluted before serving. RTS are made of fruits like mango, papaya, pineapple, lychee etc.

16. **Squash:** Squashes have 25% fruit juice or pulp, 45% TSS, 1% acidity and 600 ppm Sodium Benzoate. It is diluted before serving. Fruits such as strawberry, grape, pineapple etc. are used to make squashes.
17. **Cordial:** Cordial is clear, sparkling, sweetened fruit juice devoid of any pulp. It contains 25% juice, 30% TSS, 1.5% acid and 350 ppm preservative. Citrus fruits like lime and lemon are used to make fruit cordial.
18. **Pigment extraction:** Carotene is extracted from carrot, lycopene from tomato, anthocynanin from beetroot are extracted to add natural colours to any kind of preserves, wherever required.